Sound Business

For a complete list of Management Books 2000 titles,
visit our web-site on http://www.mb2000.com

Sound Business

Julian Treasure

FOR SWAN, BEN AND ALICE

First edition published in 2007 by Management Books 2000 Ltd
This new edition published in 2011 by Management Books 2000 Ltd
Forge House, Limes Road
Kemble, Cirencester
Gloucestershire, GL7 6AD, UK
tel: 0044 (0) 1285 771441 fax: 0044 (0) 1285 771055
email: info@mb2000.com   website: www.mb2000.com

Printed and bound in Great Britain by 4edge Ltd of Hockley, Essex. www.4edge.com

British Library Cataloguing in Publication Data is available

ISBN 9781852526689

Contents

Acknowledgments

So many people have given me inspiration, understanding, support and the benefit of their great knowledge. This short acknowledgment seems inadequate in many cases, and there are going to be people I miss, to whom I apologise humbly. However, some people are impossible to forget, starting with my friend of 30 years and my colleague at The Sound Agency, Sid Wells, who has kept me going through thick and thin, and who (as is his wont) diligently read this book and gave me careful, considered and always valuable feedback.

In creating the second edition I have received help and advice from many people. My research assistant Emma Quayle helped me to update all aspects of the book, while several people took the time to review the first edition and suggest changes or additions: in particular I want to thank my colleagues Paul Weir and Andy Farnell, along with several friends and competitors from the fast-growing sound branding industry, especially Michael Sonne (thanks to Karsten Kjems for his generosity in giving Michael the time for such diligent feedback), and also John Groves and Alex Tsisserev.

More generally I'm grateful to the many sound and sensory experts and exponents I have worked with or drawn inspiration from over the years, including: Paul Weir, generative music and digital sound guru; Jamie Perera, composer and tireless colleague; Charlie Morrow, sonic pioneer; Joshua Leeds, passionate proponent of healthy sound; Tim and Pete Cole, inventors of Koan and godfathers of generative music; David Toop, author, curator, musician and sonic artist; harmonic chant genius David Hykes , who helped me find the beauty in my own voice; vocal expert Fergus McClelland; music psychologist Professor Adrian North; cross-modal pioneer Professor Charles Spence; TED's Chris Anderson and Bruno Giussani; the guys at the Audio Branding Academy; and all the musicians, sound designers and composers who have taught, shaped, nourished or inspired me – especially Sam Dodson, Dave Muddyman, Ian Hawkridge, Brian Eno and Evelyn Glennie.

I also want to thank the people who have supported and encouraged me in launching and growing a business where there was no market: Bill Wayland, Martin Durham, Rod Banner, John Brown, Richard Berman, Mike Potter, Paul Gibbons, Simon Kelly and Chris Lee; all The Sound Agency's clients over the years; and my fellow sensory branding explorers Martin Lindstrom, Simon Harrop, John Phillips and Grainne Newborough at BRAND sense agency. All of you have my appreciation.

Last, but so not least, come my beautiful wife Swan and my dear children Ben and Alice, who bore my absence for extended periods during the birth of this book and again during this reworking. Thanks for your love and support. You are the best.

Julian
Surrey, February 2011

Introduction

Sound affects. In just two words, this is the core message of this book. Aimed at managers of all kinds of organisations, from small businesses to giant corporations and Government departments, this book aims to explain exactly why and how sound affects people, and to offer the tools and the knowledge required to harness the great power of sound and hence to produce better results for your organisation.

We experience sound from shortly after our conception to our death, 24 hours a day, every day. Our ears are at work even while we sleep – they have no choice; we have no ear lids. What we are hearing has changed greatly over the last 200 years. In the modern world, and particularly in cities, most of the sound around us is man-made, and its quantity is increasing every year.

And yet here's a strange thing. Look around you, wherever you are as you read this. Aside from any plants, pretty much everything you see was carefully designed by someone: it is intentional, meant to look like that. Its shape, colour, texture, size – all are the result of conscious choices. We could never imagine making something without being concerned with all these qualities. But as you move through the rest of your day today, start to notice that almost nothing you hear has been designed by anyone. The sound around us is mainly accidental: things make those noises just because that's what they do. Road traffic, aeroplanes, trains, coffee machines, hums, buzzes, the acoustics of rooms – almost all the elements of the soundscape are unintentional, undesigned by-products of people and machines just doing what they do, of our world being the way it is.

Not only is most urban sound accidental: much of it is unpleasant, inappropriate and counterproductive. This is partly because it's unplanned and partly because we all go around pretending it doesn't exist, so there's no demand for anybody to put it right.

Let me be clear: this book is not an anti-noise polemic. It's more of a treasure map (no pun intended). The treasure, rich and vast and available to every business, comprises increased sales, happier customers, more productive staff, more effective marketing, an enhanced brand, higher profits and more valuable equity. It lies waiting in the land of applied sound.

Most businesses are shooting themselves in the foot every day with bad sound. This happens in any one of hundreds of auditory interactions with the world: the sound of the reception area; the sound of advertising; the sound of inbound and outbound telephone calls; the sound of offices; the sound of products and services being used; the sound of on-hold music... add all of them together and you get the sound of your business.

Most of the accidental and unpleasant sound I mentioned is made, directly or indirectly, by businesses that are simply unconscious of the sound they make. This is bizarre because they certainly care about people's opinions. Each year, trillions of dollars are spent on how businesses look, and almost nothing at all on how they sound. It's as if sound has no consequences – but it does. Sound affects, and what we'll be exploring in this book is why and how it does that, and how to harness its power to improve business results.

We'll cover the nature of sound, hearing and listening in **Part 1: Sounds Interesting**. We'll look at the main types of sound and assemble some powerful tools in **Part 2: Sound Affects**. And then, in **Part 3: Sound Practice**, we'll explore all the most common expressions of business in sound, from advertising to call centres, shops to offices, with examples, practical suggestions and simple rules of thumb. Sometimes a discussion may be informed by listening to a track, which you can listen to or download by visiting the website that accompanies this book at www.soundbusiness.biz. Where that's the case you'll see a symbol like this l. There is also a full track listing at the back of the book.

When we're done, I hope you'll be as excited as I am about the huge and mainly unexploited natural resource we've uncovered. What an opportunity there is for every business that starts to listen and become conscious about its sound! There's plenty of room for everyone in this vast new territory that's been lying unexplored behind us all this time – but the richest pickings will go to those who stake the first claims, as they secure substantial competitive advantage over their rivals.

Within 10 years I believe that everything in this book will be standard practice, and that we'll have experts developing deep knowledge based on intensive research in every specialised application of sound for business. This is going to happen partly because of competition on the supply side, but also because of demand. Sound is coming up the agenda, rising like a long-repressed memory. The quantity of academic research into the effects of sound is growing exponentially, and media coverage is constantly increasing. You only have to look at the sales of mobile music players to see the degree to which people want to take charge of their soundscapes; to change the default sound of modern living.

Every business is going to have to take responsibility for its sound sooner or later. My advice is to do it now, and do it whole-heartedly. Every penny you invest in sound using the principles in this book will surely generate a return many times over: you will gain market share against competitors who are reluctant to take on controlling their sound, your customers will love you for it – and the world will be a better place.

About the second edition

The core content and structure of the book remain as they were, although some material has been moved and some removed altogether; all of the book has been updated and new content has been added to reflect the many changes in knowledge, practice and experience since the original manuscript was completed in 2007; there are many new references and a whole set of case studies that show sound branding in action (drawn from my own agency and also from my friends and competitors in the industry). Instead of the CD that came with the first edition we now have a website where you can experience all the sounds mentioned and more. I'm excited with the upgrade and I hope it serves to stimulate the conscious use of sound in business still further. Many have taken up the challenge to do that since I wrote the first edition – and many more have yet to do so. This is for all of you.

Part 1

Sounds Interesting

We are the sound.
Evelyn Glennie

1.1 The nature of sound

The better we understand sound, the more effectively we can control it and make use of it. We need not struggle through school physics here, but it is important to grasp some of the basics about sound. In the process we will make some fascinating discoveries about our universe and about ourselves.

Vibration

Vibration is moving to and fro in a steady or rhythmic manner. Some events *are* vibrations, like a violin string sounding a note or the diaphragm of a loudspeaker oscillating back and forth as it plays our favourite song. Other events *create* vibrations, like a handclap, a raindrop hitting a window, or two tectonic plates ripping past each other. You don't hear the raindrop hitting the window; you hear the sound of the raindrop hitting the window – a set of vibrations that are a consequence of the event.

Because it is rhythmic and repetitive, any vibration has a **frequency**, its number of complete cycles per time period. Some familiar measurements of frequencies are beats per minute (bpm) revolutions per minute (rpm) and cycles per second (cps). Periodic, rhythmic movement with a steady frequency is fundamental to our universe, from the quantum level to the cosmic. Subatomic particles vibrate (and are of course often seen as waves rather than particles); as a result, atoms vibrate at varying frequencies depending on their composition.

Frequency is also key in the forces that conspire benignly to maintain the perfect conditions for our existence. The rotation of our planet on its axis and around the Sun are two familiar repeating cycles that combine to create both the predictability and the chaos of our existence – including the weather, day and night, the seasons, the measured passing of the years and our whole sense of time and of nature's dependability.

On a still grander scale, stars, energetic objects like pulsars and black holes, and galaxies all have their own cycles with frequencies, sometimes in our audible range*; and in the background all the time is the high-frequency hum of the cosmic microwave background (CMB), the echo of the Big Bang.**●

The leading current theory about the structure of the universe is that everything is ultimately composed of ultra-small, one-dimensional strings, all of which are vibrating at various frequencies in a space-time of 10 dimensions. In the early time of the universe these strings spawned the well-known but little understood (certainly by me) superstrings, vast strings that still exist, spanning thousands of light years – and always vibrating.

As we are all composed of atoms (and they are composed of strings), you and I are also vibrating continually. In a sense, each person can be thought of as a unique chord made up of many frequencies. Maybe we can perceive something of this at the unconscious level, which could explain how we can instantly form a liking or disliking for someone, or how we can fall in love at first sight. Much of our language reflects this kind of sensed harmony or disharmony: we talk about good vibes, being in sync, living in harmony, being in tune. There are disease models that postulate that all illness results from systemic vibrational disharmony of some kind, and that health and vibrational harmony are one and the same. There are now mainstream professional groups, such as the International Society for Music in Medicine, exploring the relationship between sound and health and generating a huge volume of research proving that sound, and in particular music, can be effective in many therapeutic ways: these include controlling pain; reducing blood pressure and tension; causing desirable changes in the endocrine system; enhancing recovery from strokes; and treating dementia, autism and other neurological conditions. Music has proved useful in many clinical situations such as surgery, dentistry and obstetrics.

Pure vibration is a somewhat two-dimensional phenomenon, with its push-pull happening in regular fashion. The real world is much more complex, as many vibrations interact and create compound wave

---

*   The University of Arizona's current model of supernova explosions finds that stars about to go supernova vibrate at around middle C just after their cores collapse, and that the huge acoustic energy is the final trigger for the explosion.

**● The website includes sound from the CMB, as well as extraordinary sounds from a black hole and two pulsars, reproduced with kind permission from www.spacesounds.com.

patterns, much like the choppy surface of a pond when several splashes have just happened in different places. There are also the more complex periodic fluctuations that we call **rhythm**; we have dozens of these in our bodies, including our heartbeat, breathing, brain waves and hormone secretion cycles. We'll discuss these in more detail later, in the section on the effects of sound.

Resonance

The next important concept to consider is **resonance**, the tendency or predisposition of a body to vibrate at a particular frequency. When you hit an empty bottle gently with a spoon, it sounds a note: this is its resonant frequency. Resonance depends on the composition of the object: by putting water in the empty bottle we will change its resonant frequency, which is how people tune bottle xylophones and bottle organs. Like vibration itself, resonance is a property shared by all bodies, from subatomic (quantum particles) to familiar (objects like musical instruments) to vast (astronomical bodies); also like vibration, resonance is a concept widely encountered in physics, engineering and mathematics. Human beings, in common with all material things, have natural resonant frequencies, possibly one major overall frequency and several minor ones for various organs and parts of the body. Hollow bodies, like those bottles or an organ pipe, tend to have a single, very pronounced resonant frequency; solid objects, like human beings, tend to have much smaller spikes of resonance across the frequency spectrum.

Entrainment and synchrony

Although we do have predispositions towards vibrating at certain frequencies, a strong external vibration can still alter any of our internal rhythms through the phenomenon known as **entrainment**, which is the tendency of two oscillating bodies to fall into **synchrony** (identical oscillation). This was made famous by Dutch scientist Christian Huygens in 1665 when he found that two clocks left ticking next to each other always ended up perfectly synchronised, despite the fact there was no physical connection between them. There are many examples of synchrony in nature, from millions of fireflies flashing in unison on the banks of rivers in South East Asia to flocks of thousands of starlings swooping and swirling like a single living organism, all changing direction at the same instant, never colliding.

Human beings exhibit entrainment and synchrony in group behaviours such as clapping, where in many cultures there is a tendency for the clapping to fall into unison, especially at the end of a performance. Li et al[1] describe this as "collective behaviors of complex multi-agent systems".

We also unconsciously synchronise in most face to face communication, as Edward T. Hall explains in his book *Beyond Culture*:

> 'When two people talk to each other their movements are synchronised. Sometimes this occurs in barely perceptible ways, when finger, eyelid (blinking), and head movements occur simultaneously in sync with specific parts of the verbal code (the words with pitches and stresses). In other cases, the whole body moves as though the two were under the control of a master choreographer. Viewing movies [of the details of human communication] in very slow motion, looking for synchrony, one realises that what we know as dance is really a slowed-down stylised version of what human beings do whenever they interact.'[2]

It's likely that our unconscious entrainment has a huge amount to do with the establishment of rapport, which of course is an essential skill in most business conversations and particularly in sales. Most commentators and experts agree that a major element of face-to-face communication is in fact nonverbal (I have seen estimates up to 75 per cent); these days most serious communication trainers go well beyond optimising what is said and work on both delivery and conscious body language, such as matching pace of speech and tone of voice and posture mirroring; some go one level further and include more subtle rhythms such as intentional synchronising of breathing. It's still likely that much of this communication synchrony is beyond our active control, since it involves matching heart rates, blinking patterns, pheromone secretions and tiny, unconscious gestures. With the possible exception of yoga adepts, human beings are simply not able take conscious charge of these.

Also beyond the control of most of us are our brainwaves, yet here, too, synchronisation is vital. György Buzsáki (Professor of Neuroscience at Rutgers University and author of the book *Rhythms of the Brain*) is an expert on the role of rhythm in neural networks and he regards synchronisation to be an integral mechanism in many brain circuits. For instance, he proposes that synchronization of neural cells of the neural network makes voltage fluctuations within the brain more economical. We synchronise in community too: researchers have shown that effective public speakers cause the brainwaves of their audiences to become synchronised with their own – in fact, the degree of synchronisation seems to be the major factor in determining whether most people rate a talk as good or not.

Human beings exhibit entrainment of longer-term rhythms too. It's often been observed that female roommates' menstrual cycles converge and become synchronous, and one would expect the same to happen with biorhythms in tight-knit groups such as soldiers in combat or teams engaged in intense project work.

If one body's oscillation is much more powerful than the other, the synchronous frequency arrived at will be very close to the original frequency of the strong oscillator. This is similar to the way we experience gravity: we and the Earth are exerting pulls on each other but the Earth is so much more massive than we are that our gravitational influence on the planet is not detectable, while its influence on us holds us on its surface, counteracting the centrifugal force that would otherwise throw us off into space. In the same way, if I drop you in a nightclub where there is fast dance music playing at high volume, the measurable effect will be all one way: your heart rate will increase.

The combined physical effects of resonance and entrainment can be very powerful, as in the famous example of the Tacoma Narrows Bridge, where increasing oscillations caused the bridge to collapse. Because the frequency of marching feet on a bridge could match the structure's resonant frequency, creating a feedback loop to make the bridge oscillate more and more wildly until it collapses, soldiers today always break step when crossing bridges.

One widely reported example of the cycle of entrainment and resonance creating unpleasant synchrony on a bridge concerned London's Millennium Bridge, a flagship 320 metre footbridge crossing the Thames from the Tate Modern gallery towards St Paul's Cathedral. Within minutes of opening, the bridge starting swaying alarmingly, making it hard for those on it to walk and creating nausea and not inconsiderable fear. The sway reached 20 centimetres from side to side and people were hanging on in panic. The bridge was quickly closed and the designers went back to the drawing board. They discovered that the bridge was transforming random footfall by its predisposition to vibrate horizontally at one cycle per second – exactly half the typical frequency of human footfall. The strength of this resonance was felt by the people on the bridge, who all tended to fall into step with it, and so a self-reinforcing feedback loop commenced. There was no danger of this bridge falling down, but it was practically unusable. Despite all their computer modelling and understanding of resonance, this effect had not been foreseen by the designers, who had to carry out months of expensive remedial work to remove the resonance at the problematic frequency.

This is a fascinating subject in its own right, and the best investigation of it that I have seen is Steven Strogatz's excellent book, simply called *Sync*.[3]

As we'll see later, entrainment is a very powerful force in practice, operating whenever music is deployed in public places: it can cause people to eat more or less quickly, alter their speed of shopping, and make them spend more or less money in a shop. Entrainment should be considered in every soundscape we create but it rarely is, which is why we encounter so many instances where sound, through entrainment, is doing the opposite of what would benefit customer and vendor alike.

Entrainment of human vibration has very wide applications. As we've already noted, there is a great deal of work going on in the medical arena, both mainstream and 'alternative', to explore the ways in which our bodies respond to external vibration, usually in the form of tones or music. Some of the results coming from the investigations even by the traditional healthcare sector are spectacular: sound has been shown to displace anaesthetic in operating theatres, to ameliorate the symptoms of various nervous and mental disorders, and to reduce recovery times from a variety of conditions including sports injuries.

We won't venture further into the huge field of vibration, sound and health here, but the web and the CAIRSS resource (see 'Further explorations' at the end of this book) will show the way. My TED talk on Sound Health, and the supporting post on my blog that gives many references, are also good starting points.

Sound

We will define sound as *audible vibration conducted through a medium*. We will not at this point go into the philosophical debate about the sound of a tree falling in a forest with nobody there; for now, let's agree that anything with the potential to be heard is sound.

Sound is thus one particular set of vibrations, a selection from the huge superset of all vibrations. We will never hear the notes of electrons or planets for two reasons: first, their frequencies are either too high or too low for our ears to perceive; and second, they are not conducted through a medium. We can record their vibrations and pitch-shift them into the audible range, which is how the 'sounds' of black holes, pulsars and the like on the book's website were made.

Surprisingly, according to the current thinking of the physicists studying the Big Bang, sound existed before light. In the first 380,000 years of its existence the universe was an opaque plasma of photons, electrons and baryons. There was no perceivable light because matter and energy were one, and all the photons were bound up in the plasma; it wasn't until the moment of decoupling when the expanding plasma cloud cooled to 3,000 degrees Kelvin that the photons were released and light came into existence. But there was certainly sound before decoupling because the plasma was a medium and there was plenty of vibration going on as the universe expanded unimaginably quickly, but not completely evenly.[4] Much of this vibration would have been outside the range of human hearing, but if we had been there, somehow protected from the superheated plasma, we would have heard some extraordinary sounds: this was the music of the birth of the universe.

It's interesting to note that almost all of the world's spiritual paths reflect this actual physical sequence (first sound, then light) in their creation stories, even though the scientists have only just discovered it. The Old Testament has the heavens and the earth formless, empty and dark with the spirit of God hovering (alternative translation: vibrating) over them – and only then does God say: "Let there be light." The New Testament says: "In the beginning was the word." The Hindus say "Nada Brahma", one meaning of which is "the universe is sound." The mystics of Islam, the Sufis, say that all form manifests from sound.

Looking further afield, the degree of consistency becomes quite impressive, with sound being placed at the centre of creation by religious traditions from all corners of the globe including Aztec, Eskimo, Persian, Indian, Malayan, Ancient Egyptian, Polynesian, Japanese, Chinese, Balinese, Tibetan and Ancient Greek. There is another vast and fascinating topic here but we will leave it almost untouched. For those interested in sound/music and spirituality, two great starting places are Joachim-Ernst Berendt's classic book *The World Is Sound: Nada Brahma* and *The Mysticism Of Sound And Music* by the great Sufi master Hazrat Inayat Khan.

Back in the physical world, most people know that sound always requires some sort of medium: in space, it's perfectly true that nobody can hear you scream because there can be no sound in a vacuum. The sonic medium we are most familiar with is air, though many other materials will work, often better than air.

Here's how the process of sound conduction works. Any oscillator (in other words any vibrating surface) surrounded by other matter (the medium), will transmit its vibration through that medium in all directions, like a spherical balloon being rapidly inflated and deflated with the source at its centre. The inflation stroke is called compression, and is fairly easy to grasp intuitively: if a noise happens it radiates sound energy outwards, jostling all the next-door molecules into the ones next to them, which in turn barge into their own neighbours and so on. The molecules don't actually move far: they pass on the compression like a crowd surge or a ripple in a pond following a pebble's landing.

What's not so intuitive is the recovery stroke, which is called rarefaction. Having been pushed outwards, all the affected molecules are then pulled back, so from being bunched up, they move apart. We can understand this by thinking again of familiar wave action in water: when the sea's waves break on the shore, the water advances then retreats, advances then retreats. This push-pull is the two-stroke engine that powers all sound.

Sound is far more complex than a simple concentric set of ripples in a pond, of course. First, it's in three dimensions instead of two. In a still, constant density medium, sound radiates out in a sphere, moving in all directions simultaneously. As it does so, it loses energy much more rapidly than those pond waves, because its surface area is increasing at an exponentially greater rate (the difference between the surface area of an inflating sphere and the diameter of an expanding circle) and so the finite amount of energy the sound sets off with has to spread itself more and more thinly. Mathematically this loss of perceived energy (or sound pressure) follows what is known as the inverse square law: the energy falls in proportion to the square of the distance from the source. In other words, if you move twice as far away from the source of a sound you experience one quarter of the energy; three times as far, one ninth; ten times as far, one hundredth and so on.

The second reason sound is more complex than those concentric ripples is that most sound waves are themselves complex, and there are usually many happening at once, like exploding fireworks overlapping with each other in space and time. A simple, regular ripple-type wave is known as a sine wave and it creates a sound that can be made only by a synthesiser: a pure, colourless tone with no overtones or timbre.

The graphic below shows sine waves at different frequencies. The horizontal axis is time, and the vertical axis is sound pressure. The horizontal distance from one wave peak to the next is the wavelength, and the vertical distance from a peak or trough to the axis is the wave's amplitude.

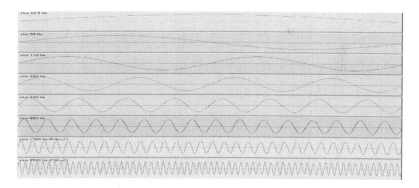

Sine waves are rarely found in the real world: all the sounds we encounter naturally have wave forms that look much more like a choppy section of the Atlantic with no easily discernable patterns present. Compare the simplicity of those sine waves with the complexity of the waveforms of four familiar sounds playing the note A at 440 Hz: a saxophone, a trumpet, a violin and a drum. The frequency is clearly the same, but the overtones create radically different timbres.*●

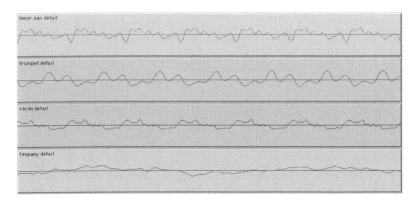

Fortunately, thanks to Fourier's brilliant work in the 19th century we now know that even the most complex waveforms can be broken down into combinations of multiple sine waves, just as a fabulous dish arises from originally simple basic ingredients. This is the mathematics that makes it possible for today's keyboards to emulate physical instruments so brilliantly. The graphic below shows the simplest start to this process, with two sine waves combining, one at low frequency (110 Hz) carrying the other, which is at much higher frequency (1.76 kHz).

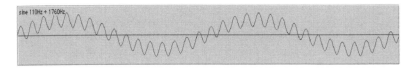

There are four key attributes that describe the envelope of a particular waveform and they are familiar to any synthesiser programmer: attack, decay, sustain and release (ADSR).

Attack, decay and release are measured in time, but sustain is a volume level. Attack is the time it takes for the sound to swell from nothing to its peak volume. For sudden sounds like percussion, or any instruments using percussive tone generation like guitars and pianos, attack is very

---

*● The website contains all the sine waves shown here, as well as the sax, trumpet, violin and drum samples to which the waveform graphs belong.

short; these sounds are sometimes called 'impulsive'. Decay is the time it takes for the sound to reduce from its peak to its sustain level, which is the volume level of the continuing sound while the key is depressed; and release is the time it takes to die away altogether when the key is released.

Although ADSR is not a complete description of a complex sound, this envelope mapping method allows sound designers to emulate most acoustic instruments – and, by applying long attack times (rarely found in the acoustic world) to create new, previously unheard sounds every day.

Harmonics

Most complex waveforms – and therefore most sounds we hear – contain overtones. Some sounds have a fairly even spread of sound energy across many frequencies, like a cymbal or rainfall, and so to us they have no particular note or 'fundamental'. But something as tuned as a violin string or a trained singer's voice contains both a fundamental note and a whole set of overtones, or partials, which tend naturally to fall at multiples of the fundamental frequency. These we call harmonics.

If you play a note on a guitar string tuned in concert pitch to the A above middle C, it will vibrate at 440 Hz. Stop the string at its halfway point and play it and you will double the frequency to 880 Hz, which is exactly an octave higher. Play at a third of the string's length and you step up the frequency another 440 Hz to 1320 Hz and a perfect fifth sounds; move to a quarter of the length and you get another octave at 1760 Hz; at a fifth you go up yet another 440 Hz step to 2200 Hz which is C#, a major third. These notes – the root, major third and fifth – are the components of a major triad chord; to Western ears they are consonant, sweet and harmonious. Our taste is in harmony with the mathematics of pitch.

Harmonic	Hz	Note	Sung	Interval	Var
1st	440	A4	DO	Fundamental	0
2nd	880	A5	DO	Octave	0
3rd	1320	E5	SO	Perfect fifth	2
4th	1760	A6	DO	Octave	0
5th	2200	C#6	MI	Major third	-14

By carrying on up the harmonic series, in our example adding 440 Hz each time by moving to a sixth of the string, then a seventh, and so on, you will eventually sound all the notes of the harmonic scale – though this is not the same as our modern 'equally tempered' scale, because the gaps between the harmonic notes are not constant. The first of the major differences creeps onto the bottom of our table, where you can see that the fifth harmonic is 14 cents flat.* (Five cents is normally accepted to be a just noticeable difference.)

Each note we hear played or sung actually contains many other notes. For most instruments and voices the fundamental is easily the loudest note, and the energy falls off with each successive harmonic. Harmonics are what give musical sounds their richness, and different harmonic profiles are a major component in the individual characters of instruments and voices. Sometimes the harmonics created by interactions between sounds create whole new sounds that we perceive even though they are not really there. For example, "Ravel in his *Bolero*, succeeds in tricking the ear into hearing not the original set of familiar instruments, but a single new 'virtual' sound source with a marvellous, previously unheard timbral quality."[5]

There is a vast volume of literature on the mathematics (and numerology) and the physics (and metaphysics) of harmonics, which turn up all over the place when you start looking. The mathematical laws of harmonics have been claimed to apply to everything from the orbits of the planets in our solar system (Johannes Kepler) to the periodic table of the elements (Rudolf Haase), and to correspond with proportions consistently found in nature and architecture.[6] For our purposes, we will rather more mundanely note that in buildings, harmonics can often cause unexpected problems. An air-conditioning system may be designed to create a hum in the bass register, too low to be bothersome to the occupants, but a higher harmonic may match the natural resonance of a duct, pipe or other structure and set up an audible hum that pollutes a room or even the whole building.

---

*   A cent is $1/100^{th}$ of a semitone in the equally tempered scale.

Soundscapes

Individual sounds are generally the province of sound designers and engineers. In our daily lives we rarely encounter one sound in isolation; usually there are multiple sounds firing off all around us. We will be using the word **soundscape** throughout this book to describe the entirety of the sound in any one location. The word was coined by Canadian composer and author R. Murray Schafer.[7] His concept of a soundscape was essentially an auditory landscape, almost exclusively applied to outdoor locations, and has been used by a thriving aural ecology movement ever since in their campaign against encroaching urban noise and their passionate efforts to record disappearing soundscapes. I hope that in the future there will be more and more recording and archiving of some of the soundscapes we're going to lose. The Internet makes it possible for virtual soundscape museums to be set up, and the effort will be extremely valuable for generations to come[*]. Each great city needs a soundscape archive because it's usually not until something has disappeared that we miss it. In London, some of the classic sounds my parents knew no longer exist and it would be fascinating to hear them: examples include the sound of tugboat whistles on the Thames, rag-and-bone men calling from their horse-drawn carts and the sound of steam trains in the great metropolitan termini. Characteristic London sounds I know so well and take for granted, like "Mind the gap" on the tube or the sound of black cabs, will not last forever.

The concept of a soundscape is complex if we start to consider it carefully. As Augoyard notes in the ground breaking book *Sonic Experience*: "The concept of 'soundscape' is 'concerned with the quality of listening [and of] what is perceptible as an aesthetic unit in a sound milieu."[8] In other words a soundscape is a relationship between the listener (with all of his or her filters), a specific space (with its acoustic properties), and an interacting set of what Augoyard calls 'sound objects'.

In his excellent book *Spaces Speak, Are You Listening?* Barry Blesser

---

[*]   The British Library UK Soundmap (http://sounds.bl.uk/uksoundmap/index.aspx) brilliantly uses crowdsourcing by inviting anyone with the excellent AudioBoo app to record and upload content. In this way the Soundmap is becoming a rich, deep and varied record with a level of local knowledge and detail that could never be achieved by a centrally created library. Launched midway through 2010, it passed 1,000 contributions within six months and continues to grow deeper and broader in coverage. This is a brilliant example of aural ecology at its best.

strongly makes the point that any building or room possesses a unique 'aural architecture', which interacts with sound sources to create the soundscape. "Just as a light is required to illuminate visual architecture, so are sonic sources to 'illuminate' the aural."[9] Sadly, most architects design only for eyes, so the aural architecture they create is often unintentional and unpleasant. Blesser distinguishes four spheres of acoustic space: the intimate sphere, reserved for friends and relatives; the personal sphere, for acquaintances; the conversational sphere, for talking with strangers; and the public sphere, which is impersonal and anonymous, determined by the acoustic horizon*. He notes that, such is the cacophony of public spaces, more and more people are designing and living inside their own personal space through the use of iPods and the like, so that "everyone lives within his or her own isolation chamber."[10] **

For our purposes we will stay out of the phenomenological complexities that create our sonic experience and simply consider a soundscape to be all of the sound that an average person hears in a particular location.

When considering the sound in a soundscape we can usefully distinguish **background sound** from **foreground sound**. This is not a hard and fast distinction because a soundscape is a continuum, but the concept is useful when designing soundscapes. Background sound (also called ambient sound) tends to be quieter, easier to ignore, more continuous, less variable, broader in spectrum; foreground sound tends to be louder, more intrusive, composed of recognisable events, changeable, located in particular frequencies. For example in a restaurant the background sound might comprise other patrons talking, the clatter of cutlery and low-level background music; the foreground sound might be our companion or a waiter speaking to us and the sound of our own cutlery and crockery. In a supermarket, background sound might include people talking, beeping tills, trolley noise; foreground sound might be a staff announcement or a baby screaming right next to us. The distinction is totally situational, but when we are designing soundscapes we need to ask what foreground sound people will be trying to focus on, and what

---

*   We will be using a similar classification in Part 3 of this book when we discuss how to create productive soundscapes in various spaces.
**   This, too, we will be discussing in detail in the section on personal soundscapes – along with its implications for hearing loss.

background will be most conducive to that happening.

When we shift perspective to that of the listener, there is a close mapping of background sound onto noise, and of foreground sound onto signal. The signal to noise ratio is key in speech intelligibility, and if the noise is too high we encounter the 'cocktail party effect', where it becomes hard or even impossible to understand conversation in a room full of chatter – a challenge that grows with age and is made far worse with loud background music added to the mix.

In some soundscapes the background effectively becomes the foreground: conversation is not the primary function in a nightclub or at a football match. The question in all cases is, what's useful, appropriate and effective given the nature of the space, its function, the people in it and the brand or values behind it? We'll be looking at this in more detail in the next section of the book when we review a model I developed that helps us to analyse the effects of any sound – or to design a soundscape to produce a particular effect on the people in it.

Measuring sound

Most of the aspects of sound we've been considering are quantifiable, so it's time to define some measuring standards that we'll be using throughout the rest of this book, and that will allow you from now on to make effective judgements based on measurements of your own soundscapes.

Frequency and pitch

We've discussed frequency in general, and seen how it covers very long cycles (comets, for example, can have rotational periods around the Sun of several hundred years) and very short ones (the frequency of visible light goes as high as 790 billion cycles per second). For sound, we use cycles per second as our standard measure of frequency. These are normally called hertz (abbreviated Hz), after German physicist Heinrich Rudolf Hertz. We'll be dealing in frequencies from a few hertz to multiples of a thousand hertz (abbreviated kHz).*●

---

* ● The sine wave samples on the website are in fact a complete set of A notes doubling in frequency all the way up from 27.5 Hz to 14.08 kHz.

Pitch is human-perceived frequency. It is not always the same as pure frequency because of the presence of overtones in complex sounds; these can disguise or change our perception of the fundamental frequency. For our purposes we will ignore this subtle difference and treat pitch and frequency as interchangeable.

Pitch also describes the choices we make in assigning frequencies to notes. The modern standard is for the A above middle C to be set at 440 Hz ("concert pitch") but this was adopted only in 1939, and still varies from country to country. The intervals between notes in our 12-note scale have varied over time, too. Today in the West we have the convention of twelve-tone equal temperament, where the ratio between each note and the next higher note is equal at 1.05946. This was not always the case, and as we've seen bears only passing relation to natural harmonics in some cases. Many tuning scales have been used over time, and in fact the way we hear most classical music today (certainly most live performances) is slightly different from the way it would have been heard when first performed, because in the time of Bach or Beethoven the instruments were tuned using a 'well-tempered' system rather than equal temperament.

Sound pressure level and loudness

Any vibration has not only frequency, but also amplitude. Going back to our simple sine wave, amplitude is the distance between the peak and the trough of the wave. It reflects the energy in the wave. Thus, in general, larger amplitudes create louder sounds. For our purposes we'll leave amplitude in the physics textbooks and instead concentrate on its direct relation, which is sound pressure level (SPL).

SPL is physical and objective, and is not to be confused with loudness, which is perceptual and subjective. The standard measure of SPL is decibels (dB) and we'll be using these quite often, so it's useful to take a moment to understand them.

The range of audible sound intensities is huge: incredibly, the loudest sound we can tolerate is one trillion as powerful as the quietest sound we can detect! Thus, in order to be manageable, the decibel scale is logarithmic. Zero dB is defined as the threshold of hearing, usually exemplified as the sound of a mosquito flying three metres, or 10 feet, away. Each increase of 3 dB represents a doubling of the power of a sound, but because our hearing also operates logarithmically (more on that later) it takes an increase of 10 dB for us to perceive a doubling in volume. This is a valuable number to remember, as are the following generalised benchmarks: 20 dB is a silent room; 40 dB is a whisper; 60 dB is a ordinary conversation or a typical office or restaurant; 80 dB is shouting or a busy street and the level at which prolonged exposure can damage hearing; 85 dB is the level at which, if sustained, employers must offer staff hearing protection by law in Europe; and at a sustained level of 90 dB hearing protection must be worn by law.

The loudest sound ever reliably recorded for accurate calibration was the Krakatoa eruption in 1883, which is estimated to have been the equivalent of 310 dB at one metre (though at that range death would have been instantaneous from the sound alone); it was heard 3,000 miles away and even at 100 miles distant it was impossible to hear someone shouting into your ear!

Sounds like that fortunately don't happen very often. Most of the time we'll be considering SPL in the range of human experience and tolerance, which is up to about 120 dB – the equivalent of sitting in the front row of a major rock concert and, for most people, well into the realm of physical discomfort.

The comprehensive table of sound pressure levels below owes a great debt to William Hamby of www.makeitlouder.com as it includes many items found in his excellent Ultimate Sound Pressure Level table.* Our table combines reference readings (taken with a sound pressure level meter) with mathematically calculated estimated levels where readings are impossible – for example with explosions or inside a tornado – and also some averages for typical sounds that may vary. I have added some typical levels for common soundscapes to complete the picture.**

dB	TYPE	DESCRIPTION
0	N	BEGINNING OF HEARING – A MOSQUITO 10 FEET AWAY. THE EAR DRUM MOVES LESS THAN 1/100 THE LENGTH OF AN AIR MOLECULE
10	P	ABSOLUTE SILENCE, ATT-BELL LABORATORY "QUIET ROOM"

---

*   William's original table can be found at www.makeitlouder.com/Decibel%20Level%20Chart.
txt
**   All dB levels are calibrated at one metre distance unless stated

13	P	ORDINARY LIGHT BULB HUM
15	N	A PIN DROP FROM A HEIGHT OF 1 CENTIMETRE AT A DISTANCE OF 1 METRE
20	A	EMPTY CONCERT HALL
30	P	QUIET NIGHT TIME IN DESERT; QUIET BEDROOM AT NIGHT
40	N	WHISPER
50	A	TYPICAL OFFICE NOISE LEVEL
40-60	A	NORMAL CONVERSATION
70	A	CAR INTERIOR
80	A	HEAVY TRAFFIC AT 10M
85	A	BEGINNING OF HEARING DAMAGE, EARPLUGS SHOULD BE WORN
90	A	HEAVY TRUCK AT 10M
100	A	TYPICAL CAR OR HOUSE STEREO AT MAXIMUM VOLUME; A LOUD SHOUT
104	P	BEGINNING OF PAIN AT THE MOST SENSITIVE FREQUENCY OF 2750 HERTZ
110	A	NIGHT CLUB ON THE DANCE FLOOR
107-114	P	VERY LARGE, POWERFUL PORTABLE RADIO
116	A	HUMAN BODY BEGINS TO PERCIEVE VIBRATION IN THE LOW FREQUENCIES
120-130	N	FRONT ROW AT A ROCK CONCERT
125	A	DRUM SET AT THE MOMENT OF STRIKING (CONTINUOUS LEVEL 115DB)
126-130	A	TYPICAL PROFESSIONAL DJ SYSTEM
128	P	LOUDEST HUMAN SCREAM MEASURED AT A DISTANCE OF 8 FEET 2 INCHES
128	A	HUMAN HEAD HAIR BEGINS TO DETECT VIBRATION; HUMAN CAN BEGIN TO DETECT VERY SLOW "BLAST WIND" OF 0.124 METRES/SECOND
132	A	EARDRUM "FLEX" TOTALLY NOTICEABLE
133	N	GUNSHOT AT EAR LEVEL, MAY VARY GREATLY TO SIZE AND TYPE OF GUN, DURATION CONVERTED TO ONE SECOND (PEAK LEVEL MAY REACH 140-160)
130-135	N	LARGE TRAIN HORN
140	A	ALL FREQUENCIES ARE PAINFUL; EXTREMELY DAMAGING TO HEARING NO MATTER HOW SHORT THE TIME EXPOSURE; HUMAN THROAT AND VOCAL CORD VIBRATION BEGINS
141	A	HUMAN BODY BEGINS TO FEEL NASUEA AFTER A FEW MINUTES
144-145	A	HUMAN NOSE ITCHES; VISION BEGINS TO VIBRATE MAKING IT SLIGHTLY BLURRY
147	N	FORMULA 1 RACING CAR
148-149	A	HUMAN VIBRATION VERY UNCOMFORTABLE AND SLIGHTLY PAINFUL; HUMAN LUNGS AND BREATHING BEGINS VIBRATING TO THE SOUND

150	N	ROCK CONCERT "THE WHO"
152-153	A	HUMAN VIBRATION IS PAINUL AND ALSO FELT IN ALL JOINTS; THROAT IS VIBRATING SO HARD IT IS ALMOST IMPOSSIBLE TO SWALLOW
154	A	TOY BALLOON POPPING (DEPENDS ON TYPE)
160	P	FLASHLIGHTS EXHIBIT ELECTROMAGNETIC PULSING (DIMMING DURING SOUND)
162	N	US FESTIVAL ROCK CONCERT 1983 (400 KW PA)
153-163	N	NHRA DRAG RACING CARS: 5,000 TO 7,000 HORSEPOWER, LIQUID NITROGLYCERINE FUEL, EARTHSHAKING AT 50 FEET, HUMANS FIND IT HARD TO SEE AND BREATHE
163	N	WHALE SONG
164	P	INTERNAL SOUND PRESSURE OF A LARGE JET PLANE TURBINE MOTOR
145-165	NP	FIREWORKS AT PROFESSIONAL PYROTECHNIC SHOWS
165-180	A	TYPICAL THUNDER
165	N	BOEING 727 TAKING OFF
150-171	P	WORLDS LOUDEST CAR STEREOS, UP TO 80 SPEAKERS, 32 CAR BATTERIES,100 KW
172	N	BOEING 727, 737, 747, 757, 767 CRUSING AT 6 MILES HIGH MACH 0.84
172	A	FOG IS CREATED, DEPENDING ON THE TEMPERATURE, DEW POINT AND HUMIDITY; AIR BEGINS TO HEAT UP DUE TO COMPRESSION
191	N	EXPLOSION – 1 LB BOMB OR GRENADE
190-195	P	HUMAN EARDRUMS RUPTURE 50 % OF TIME
198-202	P	HUMAN DEATH FROM SOUND (SHOCK) WAVE ALONE
212	N	SONIC BOOM AVERAGE FROM JET
213	N	EXPLOSION – 1 TON OF TNT, 23 FOOT WIDE CRATER
215	N	SPACE SHUTTLE LAUNCH EXHAUST, APPROXIMATELY 3 MILES PER SECOND
215	N	THUNDER FROM THE LARGEST 'POSITIVE GIANT' LIGHTINING STRIKES
215	N	BATTLESHIP NEW JERSEY FIRING ALL NINE 16" GUNS
216	P	INSIDE A NORMAL CAR ENGINE CYLINDER WITH A 9:1 COMPRESSION RATIO
220	N	EXPLOSION – LARGEST BOMB USED IN WWII, WEIGHING 11 TONS
220	N	SATURN 5 ROCKET, MELTS CONCRETE AND BURNS GRASS ONE MILE AWAY
220	N	SPACE SHUTTLE LANDING SONIC BOOM WITH VELOCITY OF MACH 20
225	P	INSIDE A NORMAL DIESEL MOTOR SEMI-TRUCK CYLINDER 25:1 COMPRESSION
229	N	SEAFLOOR VOLCANIC ERUPTION
232	N	LARGE NON-NUCLEAR EXPLOSION, 500 TONS, 1917 DESTRUCTION OF GERMAN WWI TUNNELS IN MESSINES RIDGE BELGIUM, HEARD OR FELT IN LONDON

235	NP	EARTHQUAKE 5.0 ON RICHTER SCALE
240	N	TORNADO, FUJITSU 5, ENERGY ESTIMATE BASED ON 300 MPH WIND, 1 MILE WIDE
243	N	LARGEST NON-NUCLEAR EXPLOSION EVER, 1947 DESTRUCTION OF NAZI U-BOAT PENS (USED 7,100 TONS OF EXPLOSIVE)
278	N	NUCLEAR TEST – 15 MEGATONS, 1954, BIKINI ATOLL
282	N	LARGEST HYDROGEN BOMB EVER DETONATED (57 MEGATONS), 1961. SHOCK WAVES CIRCLED THE EARTH 3 TIMES, FIRST ORBIT TOOK 36 HR 27 MIN
286	N	MT SAINT HELENS VOLCANO ERUPTION BLEW DOWN TREES 16 MILES AWAY AND BLEW OUT SOME WINDOWS IN SEATTLE-TACOMA 200 MILES AWAY
296	N	EARTHQUAKE 8.6 ON RICHTER SCALE – GROUND MOVED UP AND DOWN 13 FEET.
302	N	TUNGUSTA SIBERIA METEOR, BLEW DOWN HOUSES 600 MILES AWAY
310	N	KRAKATOA VOLCANO ERUPTION, 1883. CRACKED ONE FOOT THICK CONCRETE AT 300 MILES; CREATED A 3,000 FOOT TSUNAMI; HEARD 3,100 MILES AWAY; SOUND PRESSURE CAUSED BAROMETERS TO FLUCTUATE WILDLY AT 100 MILES INDICATING LEVELS OF AT LEAST 170-190 DB AT THIS DISTANCE; CAUSED FOG TO APPEAR AND DISAPPEAR INSTANTLY AT HUNDREDS OF MILES; ROCKS WERE THROWN TO A HIEGHT OF 34 MILES. DUST AND DEBRIS FELL CONTINUOUSLY FOR 10 DAYS AFTER BLAST. PRODUCED VERY COLOURFUL SUNSETS FOR ONE YEAR, EJECTED 4 CUBIC MILES OF MATTER. CREATED ANTI-NODE OF NEGATIVE PRESSURE AT THE EXACT OPPOSITE SIDE OF THE EARTH. SOUND COVERED 1/10 OF THE WORLD'S SURFACE, SHOCK (SOUND) WAVES "ECHOED" AROUND THE EARTH 36 TIMES AND LASTED FOR ABOUT A MONTH!
320	N	TAMBORA INDONESIA VOLCANO ERUPTION, 1815. ESTIMATE BASED ON 36 CUBIC MILES EJECTED, APPROXIMATELY EQUAL TO 14,000 ONE MEGATON NUCLEAR BOMBS. INTERNAL PRESSURE IS BELIEVED TO BE ABOUT 47 MILLION P.S.I. = 347 DB.

Decibels have meaning only when combined with measures of frequency, distance and time. Any SPL figure must first be defined in terms of **frequency**. So far we've been discussing average SPLs over the whole frequency range, but only white noise is equal in intensity at every frequency: every sound or soundscape we encounter in the world has a profile, a waveform, with peaks at some frequencies and troughs at others, so a straight average can be a misleading measure. Humans are more sensitive at some frequencies than at others (we are particularly sensitive at around 3 kHz, but by contrast we can tolerate relatively high levels of very low or very high frequencies), and as a result it's common to weight average decibel measurements to reflect the peaks and troughs in our hearing. This is called A-weighting, and measures which are weighted this way are usually written dbA or dB(A). Most sound pressure level meters are set to measure this way by default, even though it is not experimentally proven that A-weighting is really accurate in mapping the psychoacoustic effect of sound on humans. Nevertheless the practice is widespread, so from now on all measurements in decibels averaged over frequencies will be A-weighted unless stated.

Acousticians are well aware of the uneven sensitivity of the human ear, and they prefer to be a little more specific than to take simple weighted average SPLs. They have created standards for ambient noise as sets of curves that accurately reflect our sensitivity over the frequency range. These Noise Rating (NR) or Noise Criterion (NC) curves allow a room's ambient noise level to be consistently described as, for example, NR 50 if its highest frequency-specific SPL readings match but do not exceed those on that curve. NR is an ISO standard; NC is more used in the USA.

The table below shows some recommended maximum ambient noise levels for various types of space given in all three formats – NC, NR and also dB(A) – for comparison.

Space type	NC	NR	dB(A)
Empty concert halls and theatres, recording studios	10-20	20	25-30
Bedrooms, television and radio studios, conference and lecture rooms, churches, libraries	20-23	25	25-30
Living rooms, board rooms, conference and lecture rooms, hotel bedrooms	30-40	30	30-35
Public rooms in hotels, small offices, classrooms, courtrooms	30-40	35	40-45
Drawing offices, toilets, bathrooms, reception areas, lobbies, corridors, department stores	35-45	40	45-55
Kitchens, laundry rooms, computer rooms, canteens, cafes, supermarkets, busy offices	40-50	45	45-55

Many SPL levels must also be defined in terms of **distance**. A jet engine may create over 160 dB of sound pressure internally, but when a plane with four such engines flies overhead at high altitude we hear a sound that we perceive as relatively quiet because that large amount of acoustic energy has dissipated over a huge volume of air. One metre is the standard distance we use at The Sound Agency when discussing SPL for a specific sound source like a chiller cabinet or an air-conditioning duct. We'll use it from now on in this book.

Of course measurements of whole soundscapes in spaces have no distance dimension because we're actually inside the soundscape. When I'm recording the SPL of a soundscape in a space I will generally decide on four or five representative locations in the room (for example, just inside the entrance, next to the coffee machine, in the exact centre and so on) and take readings in each of them, then create an average to reflect the overall level. I will also note the time, day and occupancy level so that if I go back and re-measure after making any changes, I will be able to capture comparable data.

Finally, we may need a **time** dimension. A child's balloon popping is very loud (up to 154 dB depending on the type) but the sound lasts only a fraction of a second. In terms of annoyance value, one such sound, although loud, is nothing compared to constant traffic noise at 80 dB at night time when trying to sleep. Measures of general ambient noise, or of long and varying sounds, are usually averaged over time for this reason. A typical report on traffic noise would use a measure called Leq(x), which is calculated by taking an average SPL in decibels over the time period x, which could be five minutes or 24 hours. When we're measuring in the field, we find that an average over 10 seconds is a good standard both for specific sound sources and for soundscapes. Unless otherwise noted, this 10-second average is what I'll be using if I mention a specific SPL.

Let's just restate that decibels do not actually measure loudness. There are in fact scales of subjective loudness derived from experimental work and measured in quantities called phons and sones, but these are little used in practice. SPL has become a standard largely because it is a consistent physical property that doesn't depend on human judgement. Please do remember, though, that in every practical application average SPL measurements are just part of the story, however carefully they are taken. The average SPL in a restaurant may look fine at 55 dB, but there may be a spike at one particularly unpleasant or resonant frequency that makes the human experience of eating there miserable; this is why acousticians are dogmatic about using their NC or NR curves, which do reflect this kind of situation. It's always best to trust your ears, and to treat simple average SPL measures as a starting point and not a definitive answer.

Acoustics

Audible sound waves range in length from around 17 metres for the lowest perceptible tones to around 1.7 centimetres for the highest. There are three main ways in which these waves can be affected by matter they encounter, whether gas, liquid or solid. We've already seen that they can be **transmitted** (passed through any matter which acts as a medium); they can also be **reflected** (bounced back) or **absorbed** (transformed into heat energy). Most materials do all three of these at once, but the proportions of each vary dramatically from material to material.

There are some more subtle physical effects that influence sound waves: they can be refracted by varying densities of air, and there is a whole class of mathematical study looking into what is known as 'interference', which focuses on what happens when two or more waves collide. We'll leave these for the specialists and concentrate on the main three: transmission, reflection and absorption.

Transmission

To be absolutely correct, sound is conducted through matter rather than transmitted, but most people prefer to use the term transmission, so we will generally follow this convention.

High levels of sound transmission are desirable if you're trying to communicate, but not so good if the snoring of the person in the next hotel room keeps you awake. Good transmitters not only carry sound a long way, they move it more quickly as well: the speed of sound is absolutely not a constant.

Through air, at sea level, sound travels about 344 metres per second, which is around 1,239 kilometres per hour. This is the famous Mach 1. However, even at sea level air is not dense enough to be the best conductor for sound: in water at sea level, sound will travel around 1,430 metres in a second (a speed of 5,150 kilometres per hour), which is more than four times faster than in air. Losing less energy as it moves more easily, sound also travels further in water, which is why whales can communicate with each other over distances of hundreds of kilometres in the right conditions – though they are less and less able to do so every year as our ships create more and more noise pollution in the ocean water, not to mention the horrifying consequences of the latest submarine active sonar systems, which deafen and cripple cetaceans over large distances.

The most efficient sound conductor of all is metal, which when struck transmits sound at over 4,800 metres per second (around 17,700 kilometres per hour) – almost 15 times faster than air.

It's a common misconception that soft materials do not conduct sound. Many people put foam or heavy fabrics on the walls or heavy carpet on the floor in order to reduce sound coming in from neighbouring rooms or properties – or in an effort to reduce outgoing disturbance. These materials will affect the acoustics inside the room because they do absorb some sound – more on this below – but they also conduct quite well so they are not very effective in reducing sound transmission. The best way to block sound is to use appropriate building materials and techniques in the first place: these include walls made of low-conductivity materials; double skins with air gaps in between (like double-glazed windows); airtight seals around openings like doors and windows (sound will exploit any tiny hole to escape, so even a keyhole makes a huge difference); and filling resonant spaces with sound-absorbing material.

Sealing a room which wasn't built to stop sound can be difficult and expensive. When acousticians construct recording studios in existing buildings they generally create entire new rooms that float inside the existing skin of the building on rubber, which is a very poor transmitter. When we are called in to help clients improve the privacy between offices, the client is usually expecting us to recommend putting some foam on the walls. In fact the walls are usually not the main problem. Office partitions typically extend only up to the height of a suspended ceiling made of cheap polystyrene tiles (which reflect and conduct sound but absorb almost none of it), so the sound simply jumps over the partition into the next room. In most cases the only effective solution is to replace the existing partitions with acoustically effective ones that extend all the way up to the real ceiling and are sealed; glass partitions similarly need replacing with full-height double glazing. Retro-fitting like this is not cheap, and it's frustrating to spend money on fixing things that could so easily have been done properly within the original budget at the time of fitting the building out. However the effects can be dramatic, not only in terms of improved privacy, which is critical in many businesses, but also in terms of improved productivity arising from better acoustics, easier concentration and focus, less stress and tiredness, and so on.

Reflection

Hard, dense and shiny surfaces are good sound reflectors. In the main, surfaces that reflect light well also reflect sound well. At The Sound Agency we have used convex and concave mirrors to control sound, focusing it on exactly the area required. The light-reflecting analogy is not perfect because sound is colour-blind, so a shiny black wall will reflect sound almost as well as a shiny white one. Also, some specialist sound-absorbing ceiling tiles do reflect light very well while they absorb almost all sound through micro-pores.

Reflection is an essential part of any soundscape, and helps us to read the properties of a space and locate ourselves and sound sources in it. We unconsciously and instantaneously analyse reflections from the fleshy part of our ears to locate the source of any sound: subtle differences in the timings of sound waves received by the brain produce this information. We constantly use sound bouncing off objects around us to give us information about our surroundings (a sense called echolocation). Other species, like bats and dolphins, use this sense to the full, finding prey or mapping their surroundings predominantly through it. We are not such masters, but nevertheless a massive amount of data about our surroundings comes from the myriad sonic reflections around us. Anyone who's ever experienced an anechoic chamber, where there are no reflections at all, can testify to how unnatural, uncomfortable and downright weird it feels to be deprived of this continuous symphony of bouncing sound.

In sound engineering and acoustics, differentiation is made between discrete reflection (echo) and the overall reflective nature of a space (reverberation), which is composed of many overlapping and repeating echoes. What you get when you shout at the opposite wall in a canyon is an echo, with a definite, regular delay and usually a series of diminishing reflections; the way an organ chord dies away in a huge cathedral is reverberation.

Echo is rare in day-to-day life, except where it's been added by a recording engineer to create a pleasing effect in music. Elvis Presley's early success was certainly enhanced by his unique recorded sound, which involved the use of heavy 'slap-back' echo on his voice, and many others like John Lennon and U2's guitarist The Edge have used echo to create a signature, highly identifiable sound. Some forms of music, such as dub reggae, are defined by their use of engineered echo. Other than this kind of studio-crafted effect, most of us encounter echo only when we visit a canyon or a large cave.

Reverberation is something very different. We encounter it any time we are in any sort of room or confined space, and it can either enhance or destroy the experience we have there. The almost perfect reflection of sound by glass, concrete and polished wood makes it particularly unfortunate that the finishing materials most favoured by architects are… glass, concrete and polished wood. This is why so many restaurants, bars, cafes and shops are virtually unbearable when full: almost all the sound is reflected back into the space, with horrible results. Because we are so used to ignoring sound, most people in such a space don't realise why they are feeling irritable, tired and just want to leave.

I found a classic example in the in-store café of one giant supermarket in Hayes, west of London. The café's floor was terracotta stone (hard floors are common in catering because they're easier to clean). The wooden chairs and tables had steel legs; every time someone moved there was an ugly screeching sound like nails on a blackboard, only much louder. With at least a hundred people in there, this noise was almost continuous. The cutlery was steel and the crockery was china, contributing a cacophony in the higher frequencies. The walls and ceiling were concrete, plasterboard or glass, and they reflected back most of the sound that hit them. The kitchen was open along one side of the café so much of the banging and clattering from the cooking area was entering the seating area. When we went in, the café was about three quarters full and the noise in there was beyond belief: screeches, bangs, crashes, clatters, children shouting, babies screaming, people bellowing at each other to be heard – and in one corner, the supermarket had kindly provided a plasma TV with its sound up, in case anyone felt short of aural stimulation. The effect was a feeling of being beaten up through the ears. I imagine that one of the purposes of the café, as well as attracting visitors to the store in the first place, was to offer a chance for rest and recuperation between spells of shopping in the store. If my experience was anything to go by, most people would want to go home after this ordeal, rather than go shopping again, with obvious financial consequences for the supermarket.*●

A bass tone with a frequency of 100 Hz has a wavelength of around 3.4 metres (remember, sound travels at around 344 metres per second in air at ground level). Any room with this very common dimension between two parallel facing walls will 'trap' sound at this frequency, bouncing it back and forth in phase and increasing its energy as a result. This is called a standing wave, and it's why a stereo system or even a speaker's voice may seem to boom in certain listening positions. Depending on the frequency and the power of the standing wave, this can be a very unpleasant experience, so acousticians habitually avoid designing any space with parallel surfaces. Walls and ceilings are usually slanted at an angle specifically to avoid standing waves. Next time you visit or see a picture of a concert hall or a recording studio, take a close look at the angles in the room and you will find very few right angles or parallel surfaces.

One special form of reflection which is often used by acousticians to improve the sound of a room is diffusion, where the surface is highly irregular and thus, as it reflects, it effectively breaks the sound waves hitting it into tiny parts, sending them in many directions at once. The result is that they lose energy and are perceived as much quieter, as the new mini-waves interfere with one another, phase cancellation takes place and the energy is diffused. 'Diffusers' are special acoustic surfaces that look a little like egg boxes; most of us have seen them in concert halls or in pictures of recording studios. Some amateur acousticians actually use real egg boxes for this purpose.

Absorption

We know from daily experience that soft, air-trapping substances like curtain, carpet and foam are primarily absorbers of sound. These porous absorbers muffle sound, absorbing mainly at higher frequencies. The lower bass frequencies with their longer wavelengths are the most penetrating, travelling further and managing to move through media that

---

*● We recorded this soundscape for posterity and it's on the website so you too can sample its restful qualities.

will not conduct higher frequencies at all. As a rule, porous absorbers will absorb sound waves with length up to four times the thickness of the layer of absorber. This is why all you can hear outside a rock venue is the thumping of the bass drum and the throbbing of the bass, and why miles away from a major airport, you still hear the booming bass frequencies of mighty jet engines at full reverse thrust after landing.

If you line a room with a hefty 5 cm of porous absorber it will soak up most sound waves down to a wavelength of 20 cm. As we know, sound travels at 340 metres per second at sea level, so a wavelength of 20 cm translates to a frequency of 1700 Hz – roughly the A two octaves above middle C. Anything lower will pass straight through – not a great return on a huge amount of material. In order to stop all bass frequencies down to the lowest audible level at 20 Hz, the porous absorber would have to be around four metres thick!

Fortunately acousticians have risen to the challenge and developed membrane, or panel, absorbers, often termed bass traps. These are specially designed to absorb lower frequencies and are built into most recording studios and concert halls to avoid standing waves and unpleasant booms.

Porous absorbers in the form of soft furnishings have long been used to make spaces sound more pleasant. Tapestries were not invented just to look good: painting the same pictures would have been far easier than embroidering them, but the improvement in the acoustics of halls with stone walls, floors and ceilings was worth the extra effort. Carpets are not just comfortable to walk on; they are also the main sound absorber in most modern rooms. When they are absent, the difference is very noticeable. I once had cause to regret taking an architect's advice to leave the floors of a new office building as bare steel (this was at the height of the fashion for funky offices with all services exposed and a raw, postmodern look). It looked great, but whenever someone with high heels or leather soles walked past it sounded like a small local war had broken out, and if more than a couple of people were talking at once it was physically intimidating. It is disturbing to speculate how many people the world over are attempting to work in places whose acoustics make it almost impossible for them to do so. We'll be looking in detail at office soundscapes later on.

In the past, most architects and designers would not have made this mistake. They knew the importance of balancing reflection and absorption in any space. Even vast buildings like cathedrals used soft materials to alter acoustics: at the entrance, hard stone and reverberant acoustics could create a sense of space, size, reverence and even awe, while towards the altar, soft wall coverings could change the feeling to that of a private, small and safe space. It is rare indeed today to find such fine appreciation of the art of acoustic design outside of specialist concert halls.

Acousticians and their terms

Acousticians are best employed when planning a room, rather than brought in later to fix errors and the effects of poor design. If you are moving into a space and about to refit, I urge you to engage an acoustician right away. The planning stage is the time to think about the acoustic properties of all the surfaces and materials – ceilings (especially ceiling tiles), walls, window and floor coverings and furnishings. Every penny of the cost of professional acoustic advice at this stage will be a wise investment, saving expensive retrofits and returned many times in increased wellbeing, productivity and satisfaction for the people occupying the space for years to come.

Let's spend a short time looking at the acoustician's art and particularly the parameters they use, to better understand how and why to ask for professional help. I use these tools all the time in auditing spaces for clients, and they should help you to spot and solve problems in your own spaces.

All of a room's surfaces, angles, distances and relationships between these quantities combine to create its unique acoustic properties. The most important for our purposes is its overall **reverberation time** (RT). As a rule of thumb, most commercial rooms (offices, retail spaces, catering spaces) should have RT less than a second, and an ideal is around 0.6 - 0.8 seconds. This is not always possible – huge spaces like major railway stations and airport terminals inevitably have much longer RTs – but for many rooms excessive RT and the resulting discomfort are just the result of design by people who are not using their ears.

One problem with a long RT is that it degrades **speech intelligibility**, because syllables echo and overlap, jumbling words and making it hard for listeners to distinguish the original signal from the reverberation.

Anyone who's tried to understand announcements delivered though the antiquated Tannoy system into the long RT of London's Waterloo station can instantly imagine what I'm describing. There is a simple acoustic measure of this effect, called the **Speech Transmission Index** (STI) or more commonly a variant called the **Rapid Speech Transmission Index** (RASTI). A score of 1 means perfect comprehension; 0 means no words understood at all; 0.6 would indicate that 60 per cent of the words were intelligible. High STI is critical in spaces where speech is the primary aim, such as meeting rooms, classrooms, lecture halls and theatres. In the UK, new educational establishments are subject to strict STI targets following the publication in 2003 of a set of compulsory standards called Building Bulletin 93 (BB93). It's likely that tough standards will be introduced for older buildings, requiring enormous outlay on remedial measures – but not before time, because children sitting at the back in a classroom with STI of 0.5 are receiving only half their education!

The other issue with long RT is that it makes rooms louder and less comfortable because reflected sound is bouncing around for longer; people in the room have to speak louder to be heard, making extra noise which in turn bounces around, and so the cycle worsens. RT will not be flat across all frequencies, but will have peaks and troughs that may need to be adjusted if they are creating unpleasant effects. For example, a peak in RT at around 2-3 kHz will make conversation much more demanding because all the voices in the room will be amplified and clash with each other. Another undesirable hot spot is 6 kHz; an RT peak there will create a harsh, tinny ambience that people find extremely unpleasant.

As well as analysing RT, an acoustician will also create a sound map of a room that charts the frequencies in each location in the space. This reveals standing waves and acoustic hotspots, and will allow for accurate installation of any remedial measures.

Some materials – for example ceilings and carpets – should be selected at least as much for their ability to absorb sound as for their aesthetics, practicality and cost. One measure of absorbency is the **Sound Absorption Coefficient** (SAC), which varies between 0 (no sound absorbed at all) and 1 (all sound absorbed) for a stipulated frequency. As this is so frequency-specific, a more widely used measure is the **Noise Reduction Coefficient** (NRC), which is simply SAC averaged over four frequencies (250 Hz, 500 Hz, 1 kHz and 2 kHz) in order to give a good general picture of a material's overall performance. A good acoustic ceiling tile will have NRC 0.8 or greater, and thick carpet on effective underlay can do as well as 0.5. Stone, concrete, metal and glass have very low NRCs – zero in the case of marble. This explains why most grand corporate receptions have such impossible acoustics: marble, glass and metal ensure that every sound created in such a space bounces around it for several seconds before fading through diffusion.

The table below shows the NRCs of some common building materials, from lowest (zero) to highest (0.95).

Material	NRC min	NRC max
Marble	0	0
Brick, painted	0	0.02
Concrete (block), painted	0.05	0.05
Fabric on Gypsum	0.05	0.05
Gypsum	0.05	0.05
Plaster	0.05	0.05
Rubber on Concrete	0.05	0.05
Brick, unpainted	0	0.05
Concrete (smooth), painted	0	0.05
Linoleum on Concrete	0	0.05
Steel	0	0.1
Glass	0.05	0.1
Drapery, light weight (10 oz.)	0.05	0.15
Wood	0.05	0.15
Cork, floor tiles (3/4" thick)	0.1	0.15
Plywood	0.1	0.15
Soundboard (1/2" thick)	0.2	0.2
Concrete (smooth), unpainted	0	0.2
Carpet, indoor-outdoor	0.15	0.2
Polyurethane Foam (1" thick, open cell, reticulated)	0.3	0.3
Seating (unoccupied), metal	0.3	0.3
Seating (unoccupied), wood	0.3	0.3
Carpet, heavy on concrete	0.2	0.3

Concrete (block), unpainted	0.05	0.35
Seating (unoccupied), leather upholstered	0.5	0.5
Drapery, medium weight (14 oz.), velour draped to half	0.55	0.55
Carpet, heavy on foam rubber	0.3	0.55
Drapery, heavy weight (18 oz.), velour draped to half	0.6	0.6
Seating (unoccupied), fabric upholstered	0.6	0.6
Cork, wall tiles (1" thick)	0.3	0.7
Fiberglass, 1" Semi-rigid	0.5	0.75
Sprayed Cellulose Fibers (1" thick on concrete)	0.5	0.75
Seating (occupied)	0.8	0.85
Fiberglass, 3-1/2" batt	0.9	0.95

Another property best considered at the design stage is privacy between rooms, which is a function of the amount of sound that's transmitted through the dividing surfaces. The standard measure of sound transmission in materials is the **Sound Transmission Class** (STC). This number is in decibels, and measures the difference between the sound energy striking one side of a barrier (such as a wall or screen) and the sound energy transmitted from the other side. For example, if an office partition has STC 20, a noise level inside the room it contains of 60 dB will be reduced to 40 dB outside it, which would be perceived as around one quarter as loud (remember, every change of 10 dB is a doubling of perceived volume). An STC of zero means that all sound passes through with no attenuation at all. Like NRC, STC is an average across frequencies, so it should be used with some care in cases where the sound to be attenuated has a pronounced spike at one frequency. STC is an important number in the building trade, as many buildings are now subject to regulations about the amount of sound they allow to pass through their internal and external walls.

Remedial treatments

All too often, the job of an acoustician is to try and fix a room that is acoustically simply not fit for purpose. This usually means trying to remove unwanted or inappropriate sound. To do this, they use the acoustics ABC: absorb – block – cover up.

Absorb

Every acoustician I have ever met spends a lot of time looking up, because the first place to investigate when you want to absorb sound is the ceiling. Often you can replace standard tiles with high SAC varieties*, or if there is no suspended ceiling you can manufacture and fit sound absorbing panels on the ceiling: great effectiveness and few will notice any visual impact.

After ceilings, walls and floors are next in line, followed at some distance by dividing panels and then other surfaces such as the undersides of tables in restaurants. This comes down to a numbers game: absorbing sound is all about coverage. The greater the square footage you can cover with sound absorbing material, the greater the effect. The acoustician's skills will lie in analysing the frequencies that are doing the damage, designing the panels to absorb them effectively, and advising on what is worth covering and what is not.

The panels may be cloth-covered frames containing special sound-absorbing material; these look something like blank canvasses and can be any colour to match the existing décor; the modern varieties can even be over-printed with artwork, graphics or photographs using dye sublimation, so that they look like standard decorative canvas panels. They can also be supplied as freestanding screens by office equipment companies or specialist makers. My only firm suggestion is that you take professional advice and have an acoustician specify and order them – it would be very upsetting to cover the space with treatments and discover they are absorbing the wrong frequencies.** ●

Block

Modern building regulations have plenty of specifications for sound transmission between one building and the next, and in some cases

---

*   Where ceilings are concerned you may also come across a measure called Ceiling Attenuation Class (CAC). This is simply the STC of a ceiling material, in other words the number of decibels by which sound is reduced when it passes through that ceiling. High CAC is good for privacy, but it needs to be combined with high NRC or you will have most of the sound bouncing back at you.

**● Our old office was treated with these panels to reduce RT. There's a before and after on the website to demonstrate what they can achieve; immediately afterwards there is a before and after from a café in a major shop. By replacing the ceiling and installing wall panels we reduced the SPL by over 10 dB and the RT by around one second; the difference in the recordings is dramatic.

between one room inside a building and the next. Most people have experienced old-fashioned, cheap buildings where you can hear every sound from the next room; there is universal agreement that this is a bad thing because we all prefer our own privacy and the feeling of control that comes from not being invaded by someone else's sound.

Inside a room without full-scale dividing walls, it's much more difficult to block sound. This is particularly true in open-plan offices. As we'll discuss in detail in Part 3, the problem with open-plan is that a line of sight (much beloved by the proponents of team-based, flexible working) is also a line of sound. Low-level partitions have almost no effect on sound transmission. As soon as they go above eye level, partitions start to have some effect, blocking direct sound transmission (though not reflections) – but as we've seen, to block sound effectively we must place a complete seal between the source and the listener. Even going up to ceiling height is ineffective if the ceiling is suspended and the sound can jump over the partition.

Most standard office partitions are poor sound blockers, with relatively low STC. This is slowly changing, and the more companies that demand high STC as a matter of course, the lower the price premium will be on effective sound-absorbing partitions.

One useful fact for those who must have line of sight in their office is that sound blocking does not require opaque dividers. There are transparent office partitions, using double-glazed glass, which offer very good STC performance while still allowing everyone to see each other.

Cover up

The most common remedial technique for rooms with low levels of privacy is to install some sort of masking sound. Some offices are actually too quiet, and when everyone is working a quiet conversation can be completely audible to everyone in the space, making privacy impossible and guaranteeing disturbance. It's actually easier to work when there is a general background hum of conversation and no words can be made out, because your own conversation is masked by the ambient sound and everyone else's merges into an easily ignored continuum – another example of stochastic sound.

Where an office lacks a natural background masking sound, it is possible to install one. I have come across a prevalent myth here: many people think that you can actually create a curtain with sound: a sonic barrier that stops sound being transmitted through the air from one side to the other. Sadly, this is impossible. Even if they are being hugely excited by a masking sound in a vertical wall, air molecules will still happily pass on a vibration from side to side. It is not physically possible, at least with our current level of science, to force them not to do that.

Masking sound can be installed as part of an entire office furniture system, or it can be retro-fitted into an existing office. It usually comprises pink noise, typically delivered through an array of loudspeakers roughly 3 metres apart above the ceiling or in lighting trays at a sound level of around NR 40, or roughly 48 dB. Its effect is to render unintelligible conversation that arrives at the listener at less than this level, without being itself a distraction because it is stochastic and unchanging, so after a few minutes we cease to notice it consciously.

I have two personal reservations about pink noise masking systems. First, I suspect that there may be a psychological cost involved in ignoring this sound; the work involved in repressing it may create low-level tiredness or stress. Pink noise is artificial stochastic sound, not natural, and though there is no research of which I'm aware about this, I have a feeling that, as with artificial light, human beings thrive in the real thing and not the man-made substitute. Second, this kind of sound is aesthetically unpleasant. It is, after all, noise.

I would prefer to use natural sound either instead of, or as well as, pink noise. We have installed birdsong in offices with excellent results. It's not as consistent a sound as pink noise, but it is aesthetically much more pleasing, and it has a positive psycho-acoustic effect as we'll see in the next section: it helps create an ideal mind-alert/body-relaxed working state. Moving water is another alternative. Either of these sounds could be used in place of pink noise, or if that doesn't provide enough masking, mixed with pink noise to create masking sound that is a positive contribution to the office environment, rather than a necessary evil.

Of course many offices don't need any intervention because they already have masking sound going on: there may be a heating, ventilation and air-conditioning (HVAC) system making a constant masking sound, or traffic outside the windows may be doing the job. Please note, however, that to be effective, masking noise must itself not be distracting. Music and radios are very inadequate as masking systems because they just introduce one distraction to cover up another.

There is some fascinating work going on in this area. In London, Future Acoustic has developed a Reactive Sound System that responds to the level of sound being generated in a space with appropriate levels and qualities of masking sound. Its sonic curtain, a physical curtain with integral loudspeakers, can render conversation between co-workers completely unintelligible without requiring walls to be built. This is a new area of acoustics and I have no doubt we will see many developments, and a lot of research, in the next few years.

However, as with all this remedial work, it is so much better to avoid the problems in the first place with effective, sensitive design, architecture, layout and interior decoration. The famous composer and sound author Murray Schafer, is forthright in his view about masking:

> "There may indeed be times when masking techniques can be useful in soundscape design but they will never succeed in rescuing the botched architecture of the present. No amount of perfumery can cover up a stinking job."[11]

1.2 Hearing

Hearing is a relationship between sound and its perceiver. When sound waves hit us, there ensues a mysterious and amazing process that takes place first physically, then electrochemically and, when considered in totality, almost miraculously.

We actually hear with our whole bodies. Ears happen to be our specialists, evolved to hear better than any other part of us, but all of us can hear: bones, tissue, organs – even our eyes hear. If in doubt about this, consider the difference between hearing your own voice as you speak and when recorded. In the first case, with the sound originating in your throat and resonating in your chest and head cavities to create your own unique voice, you are hearing not just through your ear but also through your jawbone, skull and chest organs. In the second, the ear is much more dominant because the sound comes entirely from outside of you. Actor and voice coach Fergus McClelland taught me this trick to hear how you sound to others without recording equipment: speak into the palm of your right hand with your fingers cupped and turned back towards your right ear. That's the voice everyone else hears – less bassy and more mid-weighted because it's missing the internal resonances.

A living example of hearing through the whole body is the renowned percussionist Dame Evelyn Glennie, who became profoundly deaf as a child but learned to hear without the use of her ears, first by touching resonating surfaces but now simply by listening with every part of her body. Her brilliant musicianship is based on intense, sensitive listening without ears.

Hearing is the first sense we develop, as early as 12 weeks after conception. That's long before the ears are fully formed, but as we've seen they are merely the icing on the cake. The sound we are hearing in our every cell from that early time is predominantly the steady lub-dub, lub-dub of our mother's heart, at surprisingly high volume; later we become familiar with the tones and cadences of our mother's voice too – though very filtered as in this case we hear mainly the internal resonances and little of the frequencies that dominate in the outside world: we perceive a muffled, bassy version of her voice. Although our little heart, once it forms, is beating at a much faster pace, its sound is dwarfed by the massive pulse of our mother. In these early days and weeks, regular rhythm in three-time (lub-dub-rest) is imprinted on us, which is probably why 3/4 (waltz) time is somehow more soothing than 4/4.

When we are born it takes months to learn to see the world, but our ears are already fully active, seeking safety in that familiar mother's voice, and in soothing song. Mothers of every culture on the planet sing to their babies; according to anthropologist Steven Mithen[12] this powerful instinct may be the very origin of language, via an original proto-song form of aural communication. Richard Parncutt[13] goes back further, suggesting that we learn in the womb the links between our mother's patterns of sound (and movement) and the ensuing hormonal states we experience (because we share her blood flow). Parncutt suggests that these links are stored in 'transnatal memory' and form the basis of our instinctive emotional responses to music.

From babyhood onwards, hearing is our primary warning sense, hardwired in at lizard brain level. If a twig snaps behind you in a forest there is nothing you can do to stop from spinning around unless you are highly trained in combat: the survival instinct is too powerful. If I want to warn you that someone is about to attack you from behind, I don't wave: I shout.

Hearing goes very deep, very fast. It's also mechanically miraculous when you start to examine those special hearing organs, the ears. When sound waves strike us, some of them are gathered by those strange skin flaps on either side of our head (each technically termed an auricle, or pinna) and channelled into our auditory canals. The auricles are designed to amplify sound in the typical speech range (roughly 1 to 7 kHz), improving the signal to noise ratio by up to 20 dB.

After a short journey down the auditory canal, the gathered sound waves bump into the eardrum. This thin and delicate membrane is so sensitive it can almost detect individual air molecules hitting it. The faintest sound we can hear moves the eardrum just two atomic diameters.

In Heath Robinson style, one thing now leads to another. The eardrum translates all of the complex waveforms travelling down the auditory canals – entire symphony orchestras or the cacophony of the dawn chorus of birds in a forest – into pushes and pulls on three tiny connected bones whose graphic names most schoolchildren remember: the hammer, anvil and stirrup. We are now in the middle ear.

These miniscule bones have to last a whole lifetime, vibrating back and forth without rest (we never experience complete silence, so they never get to take a break) at frequencies up to many thousands of times every second. They are a miniature miracle of engineering. Their function is to amplify the tiny movements of the eardrum and to pass on all the information to the next stage.

This happens as the stirrup pushes against the oval window, the tiny membrane that connects these three little linking bones to the inner ear and to the real boiler room of our hearing: the cochlea. This coiled organ is filled with perilymph fluid; inside it is the basilar membrane, which is covered with around 16,000 tiny hair cells of varying lengths, divided into outer hairs and inner hairs.

Clumped together in bunches, the inner hairs are our frequency detectors: the shortest cells detect the highest frequencies, and the longest cells detect the lowest frequencies. The outer hairs protect and help sharpen the frequency response of the inner ones. The movement needed to make a hair activate is tiny, roughly the diameter of a hydrogen atom.

So far, so mechanical, albeit amazing: subtle variations in air pressure have been translated into various kinds of mechanical motion, ending with fluid washing against tiny frequency-detecting hairs. Somehow this carries all the rich subtleties of the sound around us.

At this point we move across the electro-mechanical divide. Each inner hair cell translates the mechanical energy into electrical signals by opening and closing some of their ion channels (up to 100 for each hair). Positively-charged potassium ions flowing through these channels create the electrical impulses that travel via the auditory nerve from the cochlea to the brain.

The brain's sound processing centre receives the two continuous signals, one from each ear, each slightly different due to the six-inch positional difference and the 180 degree variation in orientation between our ears. It synthesises them to create a three-dimensional soundscape: with our eyes closed, we can pinpoint the location of a buzzing fly as it zigzags around us. It listens for cues and familiar sounds or patterns, and sends appropriate commands to the rest of the body so that our physiology responds appropriately, whether at the level of basic survival (as in the twig-snapping example) or of sublime higher functions (such as complex emotional responses to music – a process that is still little understood). It stores every sound it hears as memory, so that we can recognise it in future and trim our response to be ever more effective. Hearing sounds we have stored deep in our memories, for example a nursery clock chime or a long-dead friend's voice, can evoke hugely powerful responses in us.

Our brain also does a huge amount of filtering so that we can understand the world around us. It's working all the time to select signal from noise, so that we can understand the person talking to us in a noisy room full of other people talking. We would have trouble staying sane if all sounds were paid equal attention by our brains: life would be a constant cacophony and impossible to understand. Some sounds require more work than others to ignore.

All of this goes on 24 hours a day, every day of our lives. Hearing never rests. We don't have earlids to close; even when we are asleep, our hearing maintains its tireless vigil so that we don't miss our alarm call or a sound that might indicate danger to our unconscious body. Hearing is not only always on: it is also incredible sensitive.

- In the time dimension, we can distinguish sounds just 30 millionths of a second apart.
- Human hearing can perceive a huge range of intensity, with a dynamic range of 130dB. A sound that causes permanent damage with short exposure (like a train horn at one metre, which at 130 dB will perforate eardrums) has a trillion times more power than the quietest sound the average healthy person can hear (a mosquito flying away at 3 metres, or 0 dB). In contrast, the eye's dynamic range (which can also be measured in dB) is just 90 dB[14], so in intensity our visual range is 10,000 times smaller than our aural range.*

---

* Bels measure any ratio of two quantities, so they can be applied to vision as easily as to sound. One bel (10 decibels) is an increase of 10 to the power of 1. So the difference between aural range (130dB = 13 bels) and visual range (90 decibels = 9 bels) is 4 bels, which is 10 to the power of 4, or 10,000.

- In terms of pitch, an average person with good hearing can detect from around 20 Hz to 16 kHz. That's about eight octaves. (Many animals can hear far higher frequencies; dogs and bats are two well-known examples.) In contrast, our visual spectrum is from about 405 THz (red) to about 790 THz (violet), which is a ratio of just under 2:1 – just one octave.

Hearing appears to operate logarithmically. Although an increase of 3 dB is in fact a doubling of the power of a sound, it takes an increase of around 10 dB for us to perceive a doubling of loudness. Similarly, we perceive a constant relationship that we call an octave every time frequency doubles: the difference between the note A at 220 Hz and the A above it at 440 Hz is 220 Hz, while the difference between A at 1760 Hz and the A above at 3520 Hz is 1760 Hz; the second gap is eight times larger than the first, but to us these differences sound the same. We seem to sense ratios, not differences.

As Joachim-Ernst Berendt points out, hearing is unique among our senses in that it simultaneously perceives value (a note) and relation (an interval). We can't see when a colour has a frequency exactly double that of another colour, but we can hear that proportion as an octave. Asked to sing their favourite song unaccompanied, most people will sing in very good pitch, starting at the right note: we can remember pitch as well as intervals.

It's sad that we take such a miraculous sense so much for granted, and our negligence has potentially grave consequences. The main issue is those tiny hairs in our inner ears. Like our adult teeth, these are one-time gifts: once gone, they do not get replaced. If they get bent, they can't control their ion channels properly, so they may shut down, causing deafness, or they may fire continuously, causing tinnitus. If they get uprooted, there is no reforestation programme: they are gone forever – at least with the current state of medical science – and we are condemned to permanent silence. Without them we have no sense of hearing and sound ceases to exist for us.

What terrible force can bend or destroy these precious cells? Mainly, just too much of what they exist for. Excessively loud sounds will damage these delicate hairs, creating the cumulative degradation in our hearing that we call noise-induced hearing loss (NIHL). Chemicals can also cause problems, but the major enemy is simply wear and tear: as we get older, the hairs become more brittle and less effective, even if they are undamaged by outside agencies.

Loss of high frequencies for a day or two after a loud gig or exposure to very loud sound ("temporary threshold shift") is the only warning we will get that the ice beneath our feet is getting thin. Once it breaks, there is no going back. Many long-time modern musicians suffer from serious NIHL. Phil Collins has lost 60 per cent of his hearing, while American rapper Foxy Brown has gone totally deaf. Another is The Who's lead guitarist Pete Townshend, which is why he is an active campaigner for raising awareness of the dangers of loud music among young musicians and fans alike.

Townshend says: "I have terrible hearing trouble. I have unwittingly helped to invent and refine a type of music that makes its principle proponents deaf." He blames his hearing loss not on high volume gigs, but on his decades of headphone use during recording. These days he has to take 36-hour breaks between sessions to let his ears recover. He reflects: "Hearing loss is a terrible thing because it cannot be repaired. Music is a calling for life. You can write it when you're deaf, but you can't hear it or perform it."

As we will see in detail when we consider personal soundscapes in Part 3 of this book, millions of young people are heading for NIHL right now because of listening to music on ear-bud headphones at excessive volume. Any time you can hear the music leaking out of someone's headphones, they are damaging their hearing. We could be dealing with a largely deaf generation in twenty years.

Even if we don't expose our ears to excessive noise, age takes its toll on hearing. Around a quarter of the population of the world's industrialised countries have a hearing problem, mainly due to age. By 50, around 20 per cent have some form of hearing impairment; by 70, the proportion is more than half.[15]

Hearing loss has many consequences, and none of them are good. Quite apart from the grief many people suffer when they lose music and the sound of the world, the loss of connection and communication often lead to serious mental and emotional problems, of which depression is the most common. This is a grave cost to society, and so is the loss of productivity: in most occupations, hearing well is a necessary capability.

Society's hearing is a precious resource, and it is in grave danger, mainly through the self-harm of headphone abuse, which is widely ignored because of our disregard for the whole world of sound. There are efforts to raise our consciousness of sound and to alert us to the current dangers, of which this book is one. Let's hope society heeds the alarms and people start to cherish and protect their hearing.

1.3 Listening

While hearing is a physical/electrochemical process, listening is our relationship with sound; it is an active choice and a skill. It's the art of perceiving and interpreting the sound that we experience in every moment of our lives. It's a major part of our engagement with reality; a prime sensory process connecting us with and locating us in the world around us. It's also probably the most important activity in our relationships, where we all need to feel heard, understood and valued – none of which can happen if we are not listened to.

Sadly, as Hemingway said: "Most people never listen." We discuss the reasons why we humans have become so ocular at the start of Part 2, where we examine the ways in which sound affects us all (whether we are conscious about it or not). Here we'll consider the process of listening and then discuss some tools, practices and concepts that can help us to become more conscious in our listening. Conscious listening is a new and powerful experience for most people, opening up a whole dimension of reality that had been absent, but not missed – a little like turning up the colour on a screen that had always shown only black and white. Usually I find that the very realisation that we have been unconscious in our listening is itself transformative and the ears suddenly become open. If, after reading this book, you find yourself grumbling about chiller cabinets in supermarkets or coffee machines in restaurants, be happy that you are experiencing the world more vividly than before! As long as you are conscious of sound, you have a choice to do something about it – and if that means moving away from unpleasant noise, then your wellbeing will be improved as a result.

Listening is partly a process of extraction, where we focus on a part of the sound we can hear and exclude the rest. We are surrounded by far more auditory information than we can possibly process. An area of our brain right at its base where it connects to the spinal column, known as the reticular activating system (RAS) is thought to be our editorial department.[16] The RAS decides* what we notice and what we don't; it filters the hundreds of thousands of bits of information that bombard us every day and passes on to the cerebrum only those it deems worthy of our focus. The process operates largely at unconscious level; the RAS knows what we're interested in consciously and will integrate this in its operating brief, so that a song we like or a car we're thinking about buying suddenly seems to turn up everywhere. But it also draws on our whole personality, constantly assessing raw data and deciding if we'll be interested in it because of who we are. In this way, we all create our own individual versions of reality.

Listening is also a function of memory. Language, like music, always exists in time: you can understand me when I talk to you only because you remember what I just said. On its own, each word has limited meaning: our communication would revert to very primitive levels if we had to use sentences of one word each. The flow and context are everything, and they are created by the recent words we are storing in short-term memory (along with, of course, recent nonverbal communication signals, not to mention our own preconceptions, perspectives and expectations).

I was lucky enough to see leading sensory perception expert Professor Mihaly Csikszentmihalyi, the creator of the concept of flow, speak at the TED conference in Monterey in 2004, where he confirmed that human beings have incredibly limited auditory bandwidth. Our brain is a very large storage device, like a huge and complex hard drive, but its audio input depends on our immediate short-term memory. Csikszentmihalyi's research has led him to conclude that we can handle only around seven chunks of information at once (plus or minus two). Combine this with Orme's conclusion[17] that we operate with 'attentional units' of $1/18^{th}$ of a second and you get a bandwidth estimate of 18 x 7 = 126 bits of information a second. Others have challenged the estimate of seven chunks at once, claiming our real capacity may be as low as three.[18] I know that on an average day seven seems almost superhuman to me! A capacity of three chunks at once would reduce our bandwidth down to just 54 bits per second (bps).

---

*   The RAS is generally agreed by neurologists to be a key link between our senses and the rest of us, alerting us and causing the release of the appropriate hormones to meet each situation, so it does seem likely that it is here in the brain stem that this function is happening. It certainly happens somewhere, and we'll subscribe from here on to the convention that the RAS is responsible.

One practical test is conversation. Csikszentmihalyi himself estimates that one person talking contains about 40 bps; we cannot understand two people talking at the same time because the combined bit rate simply exceeds our maximum input. This would tend to confirm the lower estimates of our audio bandwidth, and certainly puts it at less than 80 bps.

The small piece of short term memory that stores, manipulates and decodes audio events has to do the same thing for symbols, concepts and imagined words, as for example mine is doing right now as I write this. If there were someone talking next to me (not necessarily *to* me, just near me) my precious RAM would become overloaded and I would be unable to function as effectively. This is why other people's conversation (OPC) is such an irritant for so many people: it displaces our own internal voice with an uninvited and unwelcome lodger, elbowing us out of our own precious listening space. This has huge implications for people's productivity in modern office layouts; we will be looking at this in detail in the section on staff spaces in Part 3.

In order to make sense of the constant but ever-changing sound all around us, we use both our conscious and unconscious mind. There are two fundamental tricks we deploy to create useful information out of the vast streams of raw data: pattern recognition and differencing.

Pattern recognition

A great deal of listening is based on pattern recognition, most of which goes on without any conscious involvement from us. We start to recognise patterns in the womb: first our mother's heartbeat, then her voice, then possibly our father's voice and some of the sounds of the outside world (though they have to be pretty loud to be heard over the top of the heartbeat, and we hear them all in filtered form, with very reduced high frequencies). Pattern recognition is crucial for survival in most species, and not just in sound: it's how antelopes spot a predator (and why those predators try to confuse their visual patterns with camouflage); it's also how most animals pair and mate.

The most famous example of animal audio pattern recognition and response is Ivan Petrovich Pavlov's series of experiments in the 1890s, where he repeatedly sounded a tuning fork or blew a whistle just before giving dogs a meal; the learned (and involuntary) response of the dogs was to salivate every time they heard the auditory stimulus, regardless of whether food was present.

We use auditory pattern recognition all the time, and just like Pavolv's dogs we learn to link patterns with appropriate responses so deeply that our conscious mind has often barely registered the event by the time our unconscious has fully prepared us for appropriate action. A baby's cry at night will instantly release all the appropriate hormones in the mother, waking her and starting lactation without anyone consciously at the controls. The sound of our alarm clock in the morning; the note of an engine labouring before we change gear; the sound of our own ring tone in a crowded train – these all produce reactions in us, ranging from a more or less unconscious physical action (changing gear) to a full-system jolt incorporating mental, physiological and emotional changes (waking up to the alarm or, for many people, embarrassment at their ring tone sounding in a meeting and a frantic, adrenalised search for it).

Of course, the sound with the most meaning for us is the human voice, and it's there that pattern recognition reaches its full glorious potential. The same words can be said in subtly different tones of voice, or with marginally different cadences, and mean completely different things. A huge amount of our verbal communication is not what is said, but how it's said. Inflection, tone, timbre, pace, volume, pitch – we use all these and more, instinctively, as we deliver complex content through sound alone. The human voice, originating from a simple membrane in our throat and modulated by our brilliant use of head and chest resonances, plus mouth, tongue, lip and throat muscles, is the most sophisticated sound-generating instrument on the planet.

Most human languages use inflection extensively to alter the sense of words, phrases and sentences in context – though the emphasis varies, for example from the even-paced, self-possessed style of mainstream English to the passionate self-expression and sing-song inflections of romance languages like Italian. Tonal languages like Chinese rely on tone and inflection to define the meaning of words which otherwise sound and look exactly the same. Still other specialist languages, such as the whistling language of the Canary Islands, exist purely in tone and tempo, with no words at all.

It's possible that the nuancing of speech will become even more sophisticated as remote verbal communication (via mobiles) takes a greater and greater share compared to face to face verbal communication: without the nonverbal cues gathered by eyes, nose and other senses, we'll have to become better at communicating through sound alone. (That's assuming we don't end up all texting each other... but sound is so much richer than the written word, particularly the truncated version found in text, and it's so much more natural to speak and listen than to develop preternatural thumbs, that we can hope the art of conversation will survive for a while longer.)

Differencing

The other trick we use to help us make sense of the sonic world around us is differencing. Once our brain has classified a sound as continuous, or even as a repeating pattern, we simply stop paying it conscious attention; in effect we stop hearing it, though in actuality we apply a gate-like filter, sending the sound to our unconscious mind, rather than our consciousness, to listen to. We are all familiar with the surprise we feel when a noisy air-conditioning system shuts down for the night and we suddenly encounter virtual silence. For hours we've been enduring hums, whooshes and hisses and we have been completely unaware that our brains have been discarding the synaptic deliveries from our ears, a little like a diligent post room trashing junk mail as it arrives so we don't ever read it. It's only when the experience stops that we become aware of it. This is differencing: logging and being aware of differences, either in time or in a soundscape. Its origins are probably evolutionary: we have learned, like a hawk seeking a mouse from 500 metres above or a deer drinking from a waterhole in lion country, that anything that doesn't change, move or do the unexpected can usually be ignored. In modern day technology, exactly the same technique is used to compress digital video and audio: instead of recording the whole signal in binary digits, all that's digitised is a list of differences, which is much less information.

Differencing in a soundscape helps us to appreciate nuances of sound; for example, in an orchestral chord the basses help to define the violins and vice versa, just as in nature we can't appreciate mountains unless there are valleys.

Differencing in time is a huge boon to us – in fact it probably stops us from going altogether mad. Without it we would be cringingly aware of every tick and tock of a clock; of all the constant whirrs, hums and other mechanical noises we're bombarded by in modern urban life; of every one of the familiar sounds of our house; of each second of the crescendo and decrescendo as a plane flies overhead. Living would be impossible. Thanks to differencing, we are (in the main) blithely unconscious of all this redundant information, and we pay attention only to what could be important – like the snap of a twig behind us in a peaceful forest, which triggers an instant release of fight/flight hormones and a rapid and completely involuntary about-face.

One more way we make differencing even more potent as a protective filter is to treat stochastic sound as if it were constant. Let's spend a moment looking at stochastic sound, because it has great implications for designing soundscapes in the modern world.

Stochastic sound

Stochastic sound is composed of many random events, where the interval between each event is unpredictable but normally distributed. If that sounds like a statistics textbook, just think of rainfall or birdsong. You don't know when the next raindrop will hit, or the next bird will sing, but overall the sound blurs into a strangely comforting wash. One dripping tap can drive us mad; a million drops falling from the sky can lull us to sleep. Through differencing, we notice when rain starts or stops, but not that it is still steadily falling; we notice when the birds stop singing, but not that they are still singing just as they were ten minutes ago.

Stochastic sound has a special place in our listening. Ever since the dawn of our species, we've become accustomed to a soundscape composed primarily of stochastic sound. This is the soundscape of wind, water and birds (WWB). We consider this in detail in the section on natural sound, below.

Qualities of listening

Having used pattern recognition and differencing to isolate whatever we consider to be signal from noise, we then simply listen to it. Or do we? In fact, listening is far from a single, homogeneous process we apply indiscriminately to our chosen signal. There are qualities of listening, and we move between them all the time. We will isolate and examine three dimensions of listening: active-passive, critical-empathetic and reductive-expansive. We'll look at each in turn, and then we'll discuss the other option – not listening – which also has qualities we need to understand.

Active-passive

There is a great difference between listening actively, which takes constant mental effort, and listening passively (not to be confused with our typical unconscious state of just hearing and not really listening at all: passive listening is a completely conscious process in which we are fully present).

Active listening (sometimes called reflective listing) is almost exclusively used in conversation, and connects the sender and the receiver. It requires us to be highly present and to get involved: as receiver, our mind continually monitors all aspects of the listening process: the state and effects of our own perspectives and filters; all the possible intended meanings and implications of the content; the context of the sound; both parties' physical and mental state; the sender's background and the nonverbal communication they are using, and so on. Active listening is a discipline, a deliberate process something like receiving a visitor of state, where every action is intentional and formalised.

Active listening is widely used in caring professions such as therapy and counselling, where therapists are trained to reflect back the client's communication without any colouring or judgements so that the client knows and feels that they have really been heard. It's also common in education, employment interviews and journalism. It's sadly all too uncommon in the context in which it probably needs to be practised most of all, which is parenting. There are excellent parenting courses that teach it very well, for example Parent Effectiveness Training (PET)[19], but for every parent who has trained and uses these skills, there are probably hundreds who have not. The result is successive generations who grow up in continual conflict, certain only that their parents don't understand them.

Active listening is also an essential skill for sales professionals and telephone workers. Unfortunately, training courses for these people tend to focus almost exclusively on what to say, not how to say it or how to listen. There are few things more frustrating than the feeling that someone is not listening to you, but we are all familiar with feeling like that in telephone conversations with people who simply don't know how to listen properly because they've never been trained to do it. This probably results in more lost business than any other single sound-related issue.

In essence, active listening involves **intention, focus, reflection** and **summarising**. The listener must stand resolutely and consistently in the intention of understanding the speaker, as opposed to correcting, criticising or advising; they must focus all their attention on the speaker, maintaining eye contact and alert, interested but relaxed body language; and they must regularly feed back to the speaker by reflecting what has been said as accurately as possible, usually using a form of words such as: "What I heard you say is x; is that right?" Reflecting is independent of agreement: in active listening, you reflect back exactly what is said, even if you profoundly disagree or are aching to correct it or solve it. This is not easy, particularly for us men, whose typical listening style is to seek a solution and give advice. Summarizing is good habit in all lengthy communication as it ensures that both parties are in the same place. In active listening it also assures the speaker that the listener has been paying attention and has grasped the proportions of the topic.

This will be stiff and awkward at first, but with practice, active listening becomes a hugely powerful tool, and in the hands of a master it appears completely natural; the speaker simply feels a warm glow spread as he or she realises that someone is actually listening and understanding what they are saying.

Active listening has obvious applications in areas such as handling customer complaints or customer service requests, especially over the telephone, and it is equally potent in sales. Active listening establishes rapport very quickly, and rapport is the oil that lubricates any sale. It also helps sales people to properly understand the needs, fears, motivations and concerns of the people they are selling to. It's a million miles from old-fashioned high-pressure sales, where all those things are simply bulldozed out of the way to achieve a sale, usually leaving the customer feeling exploited, bruised and abused – and closing the door to any future business. The more enlightened modern business model is to open a long-term relationship built on trust, service and loyalty so that repeat business is a natural result. Active listening can be a major element in the construction and maintenance of this loyal customer relationship.

I remember in my previous business in magazine publishing, our first advertising sales person, John, was the polar opposite of the stereotyped sales executive. He was so quiet that you could never hear his side of the phone conversation. He never got excited, or pumped himself up, or did a countdown at the start of the day, or slapped himself in the face before a call: his was a gentle, quiet and polite nature. When he listened, you really felt heard – so despite the lack of energy and buzz going down the phone line, people just seemed to love buying from him. His figures were amazing. He went on to launch a division, buy it out from us, and has recently sold it for a large sum – all, I am sure, at imperceptible volume levels and without raising his voice.

If your sales and support staff are not already masters of active listening, I urge you to find the best training local to you and invest in it. The returns will be immediate and substantial.

At the other end of this scale is passive listening. Not to be confused with going unconscious in our listening and just hearing sound, this is fully conscious listening, but without any interaction, commentary or evaluation. The perfect example of this would be a Zen master meditating on the sound of a stream, completely listening but at the same time empty of conscious thought. Many people achieve a similar, though somewhat less rarefied, state of passive listening with music; this is surely one of the reasons why we find music so enchanting. Passive listening is akin to meditation, because it gives the conscious mind a rest from its habitual carousel of thoughts, self-criticisms, judgements and projections. When we are involved completely in a piece of music, we are sometimes transported to a realm where we exist for a while in peace, without our constant internal chatter.

Passive listening has a role in business, though a rather specialist one. In dealing with a very emotionally upset customer or employee, it is often best to start with a short time of passive listening, like a tree that bends in a storm, just allowing the first wave of anger or hurt to wash by. When someone is really angry, it's often counterproductive to go straight into active listening because a reflection such as: "You seem upset" can act as a red rag to a bull, eliciting the obvious reply "You bet I'm upset…" and maintaining the energy of the anger for longer. By contrast, encountering quietness, the emotion may spend itself sooner and constructive dialogue (using active listening, of course) can then begin.

Critical-empathetic

Here we are using the word critical as meaning 'involving skilful judgement as to merit', rather that in the popular idiom which implies wholly negative feedback. In critical listening we are applying conscious filters to what we hear, judging input against them and dismissing anything that doesn't meet our criteria. This has an important place in business communication. Discernment and judgement are crucial skills in any business, so this listening position, weighing the value of all the audio signals we receive, constantly sorting, comparing, selecting and discarding, is part of being effective on a daily basis. It is the default listening position for most businesspeople. One way of describing it is 'listening for the gold' – being constantly on the lookout for obvious or hidden value in what we hear. Of course it's up to us to define exactly what the gold is in each interaction.

Critical listening is vital in all meetings other than designated brainstorms. We have all sat in poorly chaired meetings where irrelevant, unstructured and inconclusive conversation wastes a lot of very expensive time. A big part of the skill of chairing a meeting is getting the whole group to apply critical listening: "Does anyone have any further actions to suggest?" or "Who's got a solution to this issue?" would both be examples of direction from the chair that provides structure, clarity and discipline in the discussion. Weak openings like "Would anyone care to comment?", or worse still a chair who just lets people butt in or start side conversations, may be coming from the empathetic end of the scale (see below) but in this context they are not useful.

Well-managed, conscious critical listening can teach great mental discipline in an office, or a family for that matter. However, the moment the meaning of critical slips from the Dr Jekyll one we're using to the Mr Hyde version ('a tendency to find fault') the whole dynamic alters. A stern parent who rips holes in every contribution from his or her children may end up with very articulate offspring (purely out of their own self-defence) but the cost in terms of lost self-worth, openness and intimacy will be incalculable. The same applies in a workplace or in a meeting: if the boss habitually takes a negative-critical listening position, sooner or later the supply of ideas and contributions will dry up as people get bored with being shot at and learn to keep their heads down.

The opposite end of this scale is empathetic listening, another valuable business skill. It's similar to active listening, though not the same. Whereas the goals of active listening are to be seen to be present and to make someone feel heard, empathetic listening's aims are to support someone and to make them feel emotionally understood. The applications of this type of listening are less in sales, more in general management and human relations (particularly motivation, correction, appraisal, personality clash, personal issues and disciplinary situations).

Communication researcher Dr Marisue Pickering of the University of Maine identifies four characteristics of empathetic listeners:[20]

1. Desire to be other-directed, rather than to project one's own feelings and ideas onto the other.

2. Desire to be non-defensive, rather than to protect the self. When the self is being protected, it is difficult to focus on another person.

3. Desire to imagine the roles, perspectives, or experiences of the other, rather than assuming they are the same as one's own.

4. Desire to listen as a receiver, not as a critic, and desire to understand the other person rather than to achieve either agreement from or change in that person.

Note that all four describe intentions. We are not required to do these things flawlessly, but to have the desire to do them. This is reassuring.

Again this kind of listening takes effort, because we naturally do project our own issues onto others, defend ourselves constantly, assume that everyone is like us, and criticise, judge and set out to change anyone who disagrees with us. Unfortunately these habits – probably learned when we weren't ourselves listened to when we were growing up – are not very useful in managing people, or in relationships in general.

Dr Pickering's research identifies ten discrete skills for empathetic listening. One or two are familiar from our discussion of active listening, but the list is so useful I make no apology for printing it in full:

1. Attending, acknowledging: providing verbal or non-verbal awareness of the other, i.e. eye contact.

2. Restating, paraphrasing: responding to person's basic verbal message.

3. Reflecting: reflecting feelings, experiences, or content that has been heard or perceived through cues.

4. Interpreting: offering a tentative interpretation about the other's feelings, desires, or meanings.

5. Summarizing, synthesising: bringing together in some way feelings and experiences; providing a focus.

6. Probing: questioning in a supportive way that requests more information or that attempts to clear up confusions.

7. Giving feedback: sharing perceptions of the other's ideas or feelings; disclosing relevant personal information.

8. Supporting: showing warmth and caring in one's own individual way.

9. Checking perceptions: finding out if interpretations and perceptions are valid and accurate.

10. Being quiet: giving the other time to think as well as to talk.

As you can see, empathetic listening goes several steps further than active listening. It involves a connection where the listener gives back something, in the form of caring, support, identification through sharing his or her own personal experience and perceptions. If we define intimacy as 'being fully known', empathetic listening involves a lot of intimacy. It requires (and builds) trust, understanding and loyalty – but it also involves some vulnerability, risk and commitment. It is powerful and effective, but is a tool to be used with care. I strongly recommend that those wishing to use it in business get themselves properly trained.

As with the whole of listening, the most important skill is to be conscious of what we are doing: to keep asking ourselves, "What kind of listening am I employing as a rule?" and, "How am I listening right now?"

Reductive - expansive

The final listening spectrum we'll consider to complete our three dimensions runs from reductive to expansive.

Reductive listening usually has a specific goal in mind. It is trying to arrive at a defined destination. Many men do this unconsciously most of the time: they are listening for 'the point', which is either something to contribute, or more often something to do to solve a problem. When they see it (or think they do) they will often interrupt in order to short-circuit the rest of the now-redundant conversation and jump straight to the pay-off. Men talking to other men habitually swap opinions, information or (by invitation only) advice, and the basic intention is to be useful. Conversation is essentially a practical matter with a purpose. Reductive listening is the norm, and the talk jumps from one goal to another, with the participants chiming in with their contributions as they see an opportunity to move things forward to the next place, whether that be group agreement on last night's football, the best route from A to B or agreeing how to repair a fridge. As each goal is achieved, there is a sense of accomplishment.

Reductive listening discards anything that doesn't move in the direction of the goal. It's similar to critical listening in that it filters incoming messages and discards some of them, but in this case the filter is not based on value or usefulness, but on direction. Unlike active listening, when deployed in sales, it hears objections simply as obstacles to be overcome, not as opportunities to understand the client's perspective and build a relationship.

I remember the following conversation I had once on a beach in Goa with a gypsy boy of about eight, who was in the business of retailing cans of cold drink and was one of the most unstoppable sales executives I have ever encountered.

> Boy: "You want Coke?"
> Me: "Not right now, thanks."
> Boy: "When you want Coke?"
> Me: "Maybe later."
> Boy: "What time later?"
> Me: "Er, well I suppose about four o'clock."
> Boy: "You want Coke four o'clock, yes?"
> Me: "Er, OK."

Sure enough, at four o'clock exactly, he was there with the can of Coke. He wasn't interested in any outcome but the one he had set as his goal for that conversation: one sale. His listening was entirely reductive, disregarding anything irrelevant and zapping objections for fun. I capitulated partly because it was very hot and a drink was not a bad idea, but mainly because I knew he would not give up until he won. I could have given him 50 objections and the result would have been the same. I emerged with a deep respect for his single mindedness and his determination – and with a slightly overpriced ice-cold Coke.

Reductive listening is powerful and effective for sure, but it can leave the speaker feeling unheard, disrespected and sore if it's used unconsciously, habitually or without consideration for others. Used selfishly, it betrays a lack of interest in what's going on for the speaker. Like all the modes of listening, it is best used consciously and even explicitly, as in: "I'm not going to listen to anything negative for the next hour." Used without that kind of careful deployment, it can alienate people who disagree with what's being said, quietening voices of dissent and creating a dangerous practice of everyone saying what they know the boss wants to hear, and no-one wanting to be the messenger who gets shot for bringing the bad news. I experienced a culture like this in a major advertising agency and it was both highly unpleasant and completely ossifying for the organisation: it soon ran into serious financial and accounting problems, simply because it had become impossible to report anything except for good news – so people just made it up.

If reductive listening is predominantly a masculine mode, expansive listening is more feminine. When women talk, they are often on a journey with no defined destination: the only objective is to enjoy the journey itself. Expansive listening is inquisitive, prepared to be surprised and delighted. It stops to smell the roses, passes the time of day, and has nowhere special to go. It notices the small details, and it changes direction on a sixpence when it spots something worth paying attention to. It dallies and it flits. Most of all, it is interested, not because it has an agenda, just because it cares.

This is why many men find women's conversation baffling and frustrating: from their perspective, the fact that it has no defined goal makes it literally pointless. It's why so many women complain that men don't listen: most men find expansive listening extremely difficult and crash in with practical solutions when a woman simply wants her problems to be heard and cared about, not solved. It's also why women in business often have to adapt in order to deal with predominantly reductive listening, refining their natural styles and modes and adopting different ways of expressing themselves that will make it through men's unconscious reductive filters.

Although it's often repressed, expansive listening is just as valuable in business as the other five listening positions. In brainstorming sessions, every contribution is agreed to be valid and nothing is discarded. This is a clear, ring-fenced instance of expansive listening. Much more importantly, expansive listening is also the best access to intuition and to inspiration for problem-solvers, for creative entrepreneurs, for product developers, writers and designers of all kinds. It's where flow originates, and where creative ideas come from. Without it, there is only fixing things. We men all need to spend more time practicing expansive listening.

Not listening

Of course there is always the other option: not listening at all. We choose not to listen when we designate something as noise instead of signal. Some illnesses, generically described as auditory processing disorder (APD) result in an inability to make sense of sound. Others, such as depression, cause people to choose not to listen, because this (along with closing our eyes) is one of the two main ways of cutting ourselves off from the world.

Sometimes people choose not to listen selectively, and over time this can give rise to form of hearing loss called stress-induced auditory dysfunction (SIAD). Many men suffer profoundly from SIAD: they become deaf at the frequencies most often used by women's voices. They have practiced not listening to their wives so often that they actually become deaf; many find it almost impossible to hear women's voices at all. This is not uncommon and it's far from a joke – I have seen it cause great distress. More generally, the same phenomenon has afflicted many people who live or work for a long time in places where there is a sound they want to shut out.

Dr Alfred Tomatis found that this same phenomenon was behind the problems some opera singers had with pitching certain notes: they could not sing what they could not hear. He designed a program using headphones, filters and beat frequencies that gives the ears a workout in the lost frequencies, and can reclaim the vacated territory and make the hearing complete again.

At least we know where we are when someone is not listening at all. Worse in many ways is distracted listening, because it often masquerades as listening when it is nothing of the sort. How many marriages, work relationships and families have foundered on the rocks of: "Are you listening to me?!" The prime modern culprit is of course TV, but distracted listening has been around as long as human beings have. One of my favourite cartoons depicts a Stone Age couple at the moment of the creation of language. She's saying to him: "We need to talk."

I believe that distracted listening is on the increase. We are being programmed by our media and by a common consensus to experience (consume?) more each day, and to pay attention in shorter and shorter chunks, as if staying focused on one thing for too long runs the risk of missing out on something else really worth having. We are encouraged to do everything faster, and to multitask so we don't waste time. We eat while we walk, listen to music while we read while we travel, talk on the phone while we walk or shop. This constant chasing would appear to older cultures bordering on psychosis. It challenges listening, turning it into an expensive luxury item that we can scarce afford to give out. It makes us less conscious in all our actions, including listening. It is also fundamentally altering our relationship with the main positive sound we have as a species contributed: music.

We consider the reasons for (and effects of) the inexorable rise of the personal stereo, in the section on personal soundscapes in Part 3; we take a look at the role of music in our soundscapes in Part 2. For now let's just reflect that most people below 30 years old listen to music most of the time while doing something else. Music has always had this role of course, providing a backdrop for other events, but never has there been so much music; never has it been afforded such wide attention by so many people; and never has the quality of the listening been so distracted.

Thanks to Brian Eno, we do have a whole class of music that's specifically designed for distracted listening. Eno created the concept of ambient music in the early 1980s with his Discreet Music series after he was hospitalised, unable to move, and left all day with a radio on, out of his reach, too quiet to listen to but too loud to ignore. He conceived of music that was written to be aural wallpaper, there if you want to attend to it, but equally successful in the background if you want to do something else. Ambient music is usually slow-paced, slow-changing and short of noticeable events – like a river, it changes and yet it stays the same.

We use these same principles at The Sound Agency in designing soundscapes for spaces, because our goal is usually not to engage people actively but to allow them to listen distractedly, or not at all.

We'll end this review of listening with two key observations. First, the most important thing about listening is to be conscious of your own quality of listening, and if possible that of those around you. This will help create a quantum improvement in your communication.

Second, whether you are listening or not, sound will have an effect on you. We now turn to look at exactly how sound affects human beings.

Part 1 References

1   Li, D Y, K Liu, Y Sun, & M C Han (2008). Emergent Computation: Virtual Reality From Disordered Clapping to Ordered Clapping. Science in China Series F: Information Sciences 51, no. 5,: doi:10.1007/s11432-008-0046-9.

2   Hall, Edward T. (1976) Beyond Culture, Garden City, NY: Anchor Books, p.72.

3   Steven Strogatz (2003) Sync, Theia Books

4   Much of this understanding has arisen from a project called Balloon Observations of Millimetric Extragalactic Radiation and Geophysics (BOOMERANG), which has gathered fine enough data for scientists to remodel the early development of the universe in greater detail than before. Releasing the new data in 2001, BOOMERANG's Italian Team leader, Paolo de Bernardis, said: "The early universe is full of sound waves compressing and rarefying matter and light, much like sound waves compress and rarefy air inside a flute or trumpet. For the first time the new data show clearly the harmonics of these waves."

5   Stephen McAdams, Emmanuel Bigand (1993) Thinking in sound: the cognitive psychology of human audition, Clarendon Press p 3

6   For a full discussion of these aspects of harmonics see Joachim-Ernst Berendt (1987) The World Is Sound Nada Brahma, Destiny Books, pp 57-91.

7   R Murray Schafer (1976) The Soundscape, Our Sonic Environment and the Tuning of the World, Destiny Books

8   Jean François Augoyard, Andra McCartney, Henry Torgue, David Paquette (2005) Sonic experience: a guide to everyday sounds, McGill-Queens University Press, p 7

9   Barry Blesser and Linda Ruth Salter, Spaces Speak, Are You Listening? Experiencing Aural Architecture (2009), MIT Press, p 15

10 Blesser, ibid, p105

11 Schafer, ibid, p224

12 Steven Mithen (2006) The Singing Neanderthals, Harvard University Press

13 Parncutt, R. (2009 b). Prenatal and infant conditioning, the mother schema, and the origins of music and religion. Musicae Scientiae, Special issue on Music and Evolution (Ed. O. Vitouch & O. Ladinig), 119-150.

14 http://en.wikipedia.org/wiki/Dynamic_range

15 Osseointegration – From Molecules to Man, (2000) published by the Institute for Applied Biotechnology, Gothenburg, Sweden, p.47

16 Ronald Bailey (1975) The Role of the Brain, Time-Life

17 John E. Orme (1969) Time, Experience, and Behavior, Iliffe Books

18 Denny C. LeCompte (1999) Seven, Plus or Minus Two, Is Too Much to Bear: Three (or Fewer) Is the Real Magic Number, Proceedings of the Human Factors and Ergonomics Society 43rd Annual Meeting, pp 289–292.

19 See www.gordontraining.com for details of PET and also the business-oriented variant Leader Effectiveness Training. The classic book is Dr Thomas Gordon (2000) Parent Effectiveness Training: The Proven Programme For Raising Responsible Children, Three Rivers Press.

20 Pickering, M (1986) Communication, published in EXPLORATIONS, A Journal of Research of the University of Maine, Vol. 3, No. 1, pp 16-19.

Part 2

Sound Affects

Hear, and your soul shall live.
Isaiah 55:3

2.1 The Emperor's naked!

Since the Industrial Revolution, noise has been on the increase. The ancient soundscape of wind, water and birds (WWB) has been progressively displaced by the sound of our machines, and in cities it has all but disappeared.

This happened without anyone really noticing, because Western culture has over the same period become dominated by the eye, with the ear relegated to a supporting role. It's interesting to ask exactly why this has happened.

There is a metaphysical dimension for certain. Sight is a sense with strong masculine aspects: it is active, outward-oriented, targeted, directed. Hearing by contrast is clearly more feminine: it is passive, receiving, always open. The imbalance between them is part of the developed countries' loss of general balance between what the Chinese call yin and yang. Yang is creative, energetic, masculine; it represents day, light, heat, force. Yin is receptive, passive, feminine; it represents night, darkness, cold and yielding. Considered in these terms it quickly becomes clear that the whole of Western culture is very yang, so it's consistent that the primary sense has been sight for so long. Even our relationship with sound has become yang: we are a culture that prefers to talk rather than to listen, to tell rather than to learn, to make noise rather than to experience silence.

There are two further, more prosaic explanations for our ocular society. First, we invented writing, from which point the premium on careful, attentive listening started to reduce. Now, with sound recording and the web, you really don't have to listen very carefully at all: if you miss it, you can always check it later. When all knowledge was passed down in oral tradition, this was not the case and the ears were the primary channel for wisdom. This caused greater sensitivity to sound in many important disciplines, including architecture, where practitioners are now almost exclusively ocular and we have all but lost a tradition of designing the sound of spaces along with the look of them. Barry Blesser, in *Spaces*

Speak, Are You Listening?, notes that "the craftsmen who actually created aural architecture… were illiterate".[21] In contrast, as Blesser succinctly puts it: "The aural architecture of many modern spaces is created by architects, space planners, and interior designers with little appreciation for the aural impact of their choices."[22]

Second, with the explosion of machines from the Industrial Revolution onwards, we have lived with ever-increasing levels of mechanical (and now electro-mechanical) noise. We have simply habituated by developing the habit of suppressing our conscious awareness of sound.

However we got here, it seems to me that in the cities of the developed nations of the modern world we are all playing a game of The Emperor's New Clothes. We stand on urban street corners, bellowing at each other over the traffic noise and pretending this is perfectly normal. If we could see or smell this amount of pollution we would never put up with it.

Our tacit societal agreement not to listen extends into many aspects of corporate life, with mainly adverse effects in all directions, including on the bottom line. We put staff in work environments that, through inappropriate sound, create great stresses on them (more of which below) and we wonder why we get phenomena like reducing productivity, lack of morale and Sick Building Syndrome.

Personally and corporately, we spend billions on how we look and almost nothing on how we sound, and we fail to notice the way in which the uncontrolled, unplanned noise we are making clashes with our carefully-constructed visual identity. For businesses this often leading to performance shortfalls in sales and marketing that we find unaccountable because we never ask questions like: "What is the sound of our brand?" or (to customers) "How do we sound to you?"

I hope that this book is the equivalent of the small boy who shouted: "The Emperor's naked!" As soon as we all recognise the mess we're making with sound, we can start to take advantage of the huge opportunity that's been right behind us all this time: to become conscious about the sound we are making, to design our soundscapes and aural communications just as carefully as we design our visual environments and communications, to make our sound consistent with our look, and ultimately to become masters of intentional sound, so that every space and every communication has appropriate, effective and pleasing sound.

2.2 The main classes of sound

The human voice

There is no sound more powerful than the human voice. Volcanic eruptions may be louder; music may be more beautiful; a lion's roar more thrilling; surf more soothing – but the human voice is the only sound that can start or stop a war, direct the course of nations, create amazing technologies, bring people together, underpin every aspect of our commercial activities, and of course say "I love you."

Our voices, like our fingerprints, are unique. There are those who whisper, those who shout; those who mutter and slur, those who boom and enunciate; there is slang or dialect, and there is correctness; there are those with musical, enchanting voices and those with flat, grating ones; there are thousands of languages and millions of accents.

Given the power of the human voice it is incredible that the vast majority of people have never had a moment's training in how to use it, and probably have never spent more than a few minutes consciously thinking about their own voice. We all learn to speak in an unconscious way, picking up from parents, school friends and other intimates our accent, our vocabulary and phrase bank, and our range of inflections and tones. Not many of us were consciously engaged in the process of developing our voice, our sound (in the sense of a jazz musician having a 'sound'). It changes over time, whether we are considering individuals or society as a whole. Listening to old recordings of people from all social backgrounds brings this home clearly: nobody today speaks the way radio announcers of the 1930s did.

And there is more than just our sound. We all instinctively manage our delivery to communicate much more than our words are saying. This metalanguage communicates our emotions, our context (background, status and so on) and our intentions, as well as altering the sense of what we say, for example with irony or sarcasm.

We all need training to become masters of our voice. Any business that invests in this kind of training will establish a major advantage in clear and greatly more effective communication. Let's look at some of the main aspects of voice that can be worked on.

Generation

Eastern mystics say that breath is life, and our voice is nothing but breath. It's no surprise then that in many traditions, the voice is our essence, our primary connection with the universe. In pretty much any tradition or society, it's the single most important manifestation of our being in the outside world.

In simple physical terms, the process of our voice starts when air is directed through our larynx, in which are located the ligaments of our vocal cords or folds. Men's cords are 17-25 mm long, while women's are 12.5-17.5 mm, which is why men's voices are deeper and women's higher. As we force air past these cords they vibrate, and we modulate this vibration with attached muscles. This creates a fundamental tone of around 100 Hz on average for men and 200 Hz for women.

Overtones

The human voice is rich with overtones up to about 3 kHz, all unconsciously created and mainly delivered consistently by each individual with their own particular signature mix. Few people with untrained voices ever consciously change their overtone profile. The best actors and singers can do this at will, however, completely changing their sound.

It is discomfiting to become conscious of one's own vocal overtones. I had this experience when training with the brilliant American overtone singer and teacher David Hykes in Denmark. Using his adaptation of Tuvan overtone singing techniques, Hykes can simultaneously sing a fundamental and also a higher-pitched harmonic, and can modulate them both separately – so he sings two melodies at once. It's stunning and captivating to hear*, but to do it yourself is like suddenly seeing the world in colour. When I first learnt how to do this, my own voice seemed suddenly to become a rich choir; car engines were singing to me, and every aural experience was infinitely richer as I could hear the harmonics in each sound.

---

*   The best introduction is the classic website Hearing Solar Winds, available from David's website www.harmonicworld.com.

Being able to sing overtones may have limited direct applications in business, but this kind of training is wonderful for team-building and it does result in a hugely heightened awareness of your own voice and the sound around us. I recommend it.

Registers

Going back to the vocal cords, we modulate their raw sound energy by using resonances in many cavities of the head, throat and chest cavity. This is what creates the different registers of the voice, and why actors and singers sound very different to untrained speakers.

There are four registers. The **chest register** produces our fundamental tone, and creates our deepest, fullest vocal sound by using the resonance of our large chest cavity. Most Africans are firmly based in this register in their speech. Because it creates much more sound energy, it's the one we need when speaking to groups, and is highly developed in actors and professional public speakers who need to project to be understood up to 100 metres away with no microphone. It's impressive to see a trained person do this without shouting. Anyone can learn, without having to have a chest like Orson Welles or Pavarotti; leading voice coach Fergus McClelland claims he can teach someone how to move their voice into this register in five minutes, though I think it may take rather longer to become a master of this skill.

The **head register** is where we generate our higher voice – the one we, in the West, use most of the time. (My wife being Italian, I know this register well: it's the one used almost exclusively by Italian women.) Resonating mainly in the cavities of the head, it's more nasal and throaty in sound than the chest register, with less bass and fewer overtones.

When we engage the **falsetto register** it feels like changing gear, and we move to a higher range of tones altogether. This is the range where Coldplay singer Chris Martin spends much of his time, and is a big part of the band's familiar and very identifiable sound. Most men rarely go here in conversation unless mimicking women's voices; however, possibly because it carries flavours of passivity, levity, even ridiculousness, this register is a defensive base for many women, who use it to stay out of the way of dominant male vocal traffic. There are vocal trainings for women in business that help them to emerge from this form of unconscious self-deprecation and move into the more assertive head and chest registers.

Finally there is the **whistle register**, little used and less understood, which creates ultra-high notes (soprano C and above). If you're a Mariah Carey fan you will be familiar with this register, but the rest of us encounter it very rarely.

Envelopes

Having generated sound and resonated it, we create envelopes to shape it by using the muscles of our mouth and face, and by directing air through our nose and mouth. Most of the content (our words) is communicated in this last step; this is why we can understand the content of a whisper, which involves no work for the vocal chords at all and is effectively just shaped noise, as well as we can the content of a passionate address. As we all know, however, it is difficult to create metalanguage in whisper. Emotion is far better conveyed when we have tone and harmonics to play with.

Projection

Being heard at the back is not just a matter of generating from the chest, though it starts there. It also requires the right breathing techniques and the right stance.

Discuss the voice with any actor and it won't be long before you hear the word 'diaphragm'. Where you and I breathe and talk from the top of our lungs, someone who is serious about projecting will be aiming to use air from right at the bottom of the lungs, and pushing it out with focused pressure from the thoracic diaphragm, the huge shelf of muscle across the bottom of our ribcage that controls our breathing. This is the secret of any big voice you've ever heard: it comes from big breath and conscious use of the diaphragm.

Almost as important is posture. It's not easy to project while sitting; almost impossible while lying down. The best posture is feet square on the ground, about shoulder width apart, facing the audience squarely and standing upright with the weight evenly balanced on the balls of the feet. The best speakers retain eye contact with the audience at all times, speaking to one person at a time and never to all of them together.

Once breathing and stance are correct, projection is all about delivery: clear enunciation, good pace and inflection (both dealt with below) will ensure that the message gets all the way to the back of even a large and crowded room.

Inflection

Another aspect of the voice that has a radical effect on our effectiveness as communicators is pitch, expressed in speech as intonation or inflection (the sing-song variation in tone we employ in order to add sense to our speech, for example raising our tone at the end of a question).

It's important to use inflection, and in business it's generally better to err on the side of too much rather than too little. Another word for boredom is monotony, which simply means 'the continuance of an unvarying tone'. If we inflict this on people, boredom is what we will be rewarded with.

Some people naturally inflect well, possibly those with a moiré musical ear, or those for whom sound is the primary sense. Others are not so naturally blessed and they simply have to work at it. Usually it isn't until we hear a recording of ourselves speaking that we get a true picture of our skill with inflection. Most of us can improve.

Inflection varies around the world, and awareness of this is important for anyone communicating with people from different cultures. Some cultures inflect a lot and use a wide range of tone changes – think of any Italian conversation – while others prefer a much narrower range and have fewer inflection conventions to choose from – for example German. Cultural differences may be small but very noticeable, for example the practice of high rising terminal (HRT), where the speaker raises his or her tone on the last syllable of a statement, as if it were a question? Once found mainly in Australia, this is now spreading rapidly among young people in the English-speaking world, particularly the UK and the US? I for one hope it dies out, as it robs speech of variety and meaning. On the other hand, differences may be large but not noticed, for example when Westerners learn tonal languages like Chinese and fail to grasp the subtleties of the intonation required in order to be understood.

Transcultural intonation is interesting and important in international business, but it is in day-to-day intonation that most can be gained. Intonation is a major element of metalanguage, and it can easily communicate key messages, such as the speaker's level of confidence, interest, enthusiasm, happiness, openness and so on. Why would we want to leave all this to chance in a business conversation? Many times I have witnessed someone whom I know to be enthusiastic and interested giving all the wrong metamessages because their unconscious intonation habits are just trundling along as normal, saying things like "I'm cool, I'm really laid back, I don't make an effort for anyone, I'm fashionably disengaged, I don't care if you like this or not because I'm not really that interested."

If I am conscious about my inflection I can give much more useful and accurate metamessages than these, such as "I'm very interested in this conversation, I'm excited, I'm confident, I'm present and fully engaged, and *this* is important." Again this is something that most people have never consciously considered, and something that is rarely trained in business. I suggest that almost any business will benefit from its people mastering conscious intonation.

Pace

How many times have you sat in some pain, listening to a highly nervous public speaker rush through a talk, jumbling words, confusing sense, randomly spraying emphasis around and succeeding only in enrolling the audience in his or her dearest wish, which is to stop talking? Or, equally painful, listening to someone whose sedate pace of delivery never varies, eventually lulling the audience through sheer monotony into a stupefied state not far from coma?

Pace is another potent tool. It can create great emphasis (by slowing down and stressing each word deliberately); it can generate excitement and enthusiasm (usually done with high-paced delivery). It's also another powerful element of metalanguage, communicating many things that we don't actually say, particularly about our own state of engagement or excitement.

Any trained top-class public speaker will use pace consciously all the way through their delivery. Why do we allow sales people, call centre staff, shop floor staff and announcers to be completely unconscious about their pace when this is such a powerful aspect of communication?

Accent

Some people are more sensitive than others to incongruity between their sound and the sound around them. I have worked many times with people who came to London from parts of the UK with strong regional accents, and who consciously worked at changing their sound, losing the regional intonation and pronunciation and replacing them with a more acceptable modern London version of what the BBC used to call 'received pronunciation'.

Accents can be a problem in business, to be sure. A very strong regional or national accent can simply get in the way of communication, making it harder for people to understand what's being said and creating scope for misunderstandings, errors and conflict. Sometimes this is unavoidable, but often it *can* be avoided either by changing the speaker or by changing the accent. It makes no sense to put someone in a role that centres on clear communication with customers when they have an accent that makes that difficult. This obvious and seemingly uncontroversial proposition immediately puts us in awkward territory because strong accents are often created by ethnicity, and it is politically incorrect – not to mention legally culpable and wrong – to deny someone a job because of their ethnicity. It would be very easy for a decision that was based on comprehensibility alone to be misconstrued as based on ethnicity and thus challenged in court, with all the resulting opprobrium. It's obviously wrong to select based on ethnicity, but at the same time it makes little sense to have station or store announcements made by someone with a strong regional or national accent unless the listeners share it – it's hard enough to hear announcements on the inadequate public address systems in most locations, without customers struggling to translate unfamiliar pronunciations and inflections. This issue has been in play recently with the ever-increasing use of overseas call centres to handle customer support calls. The Indian call centre industry has been losing ground of late to new players such as the Philippines, mainly because Indian call centres speak British English, often with a strong Indian accent, while Filipino call centres speak American English with little or no local accent. This is why the Filipino government is estimating that it will capture 50 per cent of the world's English-speaking call centre trade by 2008.

Every business should behave ethically and in accordance with local employment laws and customs of course, and equal opportunity is a rock on which better societies have been built. I hope that the legislation around these things will not prevent active selection of communicators based on how easy it is for the majority of their audience to understand them. This is surely simple common sense.

Perhaps the crucial bridge will be training. It's rare indeed to see voice or accent coaches working in businesses, but this must change. It is a great skill to be able to communicate, and another to be able to feel comfortable in many situations. The ability to adapt one's sound to suit the need at hand is one which many British teenagers have already mastered: they will speak street English dialect among themselves, complete with mock-Jamaican accent and a whole private vocabulary that the rest of us think is cool when we discover it about two years later – then in class or at home they revert to a much more standard form of English to communicate with the older generation more effectively.

In the UK, we have seen concentrations of call centres in areas which are commonly perceived as having pleasing accents, such as Liverpool, Ireland, the lowlands of Scotland and the North East. Many accents have what we could call overtones – judgements, assumptions, instant emotional reactions – which result from both the nature of the place and the people owning the accent, and the upbringing and attitudes of the person hearing it. Sometimes these overtones have a broad currency, prompting similar reactions across a wide range of listeners, hence the concentration of call centres in areas where speakers' overtones are positively perceived, and the lack of call centres in areas where speakers' overtones are negatively perceived. While a gentle Irish accent may have global overtones of honesty, light-heartedness, friendliness and a twinkle in the eye, a strong Somerset or Devon accent still, for many people in the UK, has associations with naïve, simple farming folk. Clearly, this is something of a caricature, but it is still the case that few call centres locate in Somerset or Devon and very few captains of industry or high-profile success stories speak with that accent. If they spoke with it in the past, they have long since changed their sound to adopt a more career-friendly delivery.

It's important to be sensitive to the whole subject, considering both aspects of accent – comprehensibility and overtones – when selecting communicators.

Vocabulary

I have long had a sense that the vocabulary of the average British person is shrinking, and that this has been accelerated by the growth of the Internet and the use of text and mobile phones.

This was confirmed recently by a Tesco-sponsored study by Lancaster University's Professor Tony McEnery, who found that teenagers today in the UK use only half as many words as 25-34 year olds. The top 20 words of the teenagers (which of course included yeah, no, but and like) account for a third of their speech. Their average vocabulary totalled 12,600 words, compared to the 21,400 words used by the next generation up. Professor McEnery cited technology isolation syndrome as the main culprit: this generation spends an awful lot of time playing games and listening to MP3s rather than communicating with each other. Employers are already complaining that young recruits lack the skills to answer the phone and have a professional conversation with a customer, or speak effectively to groups or in meetings. He concludes: "Kids need to get talking and develop their vocabulary."

Of course another explanation of the disparity could be that we learn a lot of words between our teen years and our mid-twenties. However this is sadly not likely to be the case, as US evidence indicates: the typical American six to 14 year old of the 1950s had a vocabulary of 25,000 words; by the year 2000 this had shrunk to just 10,000 words.[23]

In the UK we have had a major recent educational emphasis on literacy, so one hopes that reading and writing have improved. But it seems to me that there has been much less attention paid to articulacy or on its foundation, which is vocabulary. Business can help here by pressuring governments to teach public speaking in schools, and by providing training for first-jobbers in these important areas.

Mirroring

We've considered a range of tools that can help optimise the effect of the human voice in business. They all work in public speaking, and as we found when we looked at entrainment, a good speaker using these skills can cause a whole room full of people to fall into step with his or her brain waves.

All the voice skills can also be deployed where appropriate in one-to-one communication, either face-to-face or on the phone. One very effective way of doing this is mirroring, which involves consciously matching register, pace, intonation, vocabulary, even accent to those of the other person. Often this all happens instinctively to some degree, but when used intentionally it can hugely enhance the power of delivery, creating fertile soil for the messages to land in.

Mirroring needs to be approached with care and practised intensively before being tried in the field. It can come across as gauche, stiff, insincere or manipulative. It's most effective when being done in a genuine attempt to create a good connection with another person, to remove pointless obstacles that would be in the way of an important message being received.

What is extremely useful all the time is to be aware of the effect of one's own register, pace, inflection, accent and vocabulary on other people. Whether business communicators aim to mirror or not, they need to be conscious of their sound; not just doing their natural thing and being surprised when someone didn't hear them.

In effect this means listening while you are talking, which may sound impossible, like breathing in while you are breathing out, but in fact it is a hugely powerful business technique. Sadly, it is rarely taught in management training courses.

Music

Of all the types of sound, music is the one we find most fascinating. Perhaps this is because it expresses our ability to *make* sound: we can't create light with our bodies, but we can create sound. More likely it's because music affects us deeply and mysteriously; it is a language we all unconsciously speak and understand.

Music is undeniably important. It exists in every culture on Earth. It is central to mother-infant bonding and communication; it attends all our social rites of passage (comings of age, marriages, funerals); it bonds our communities (tribal dances, football chants, Band Aid); it informs and shapes our courtship and love lives (ballads, 'our song', slow dancing); it gives us courage and even strikes fear in our enemies in war (bagpipes, marching drums, jingoistic popular songs*); it shapes and focuses social change (rock'n'roll, Woodstock, punk, Live 8); it is integral in our spiritual and religious practices; and now it provides a soundtrack for many people's daily lives through personal stereos.

For hundreds of thousands of years we humans have made music. According to archaeologist Steven Mithen, music came before language. He suggests that our ancestors originally communicated via a musical, non-verbal proto-humming, which possibly originated in the instinctive music of the mother-baby relationship, and which used the metalanguage tools we have just been considering – intonation, pace, register – with no words at all. For Mithen, the advent of language sidelined music from its original role as our core communication vehicle. Language is processed by different areas of the brain to music, and these have become dominant as we have concentrated exclusively on lingual communication, leaving music as a powerful tool that we now use without really understanding.[24] The same essential argument is to be found in other recent work on music, such as Philip Ball's *The Music Instinct*. Ball writes that we are innately musical, and that learning to listen to music offers "a direct route to the core of our shared humanity".[25] Daniel Levitin, analysing the neurological effects of music in his *This Is Your Brain On Music*, also concludes that music's antiquity and ubiquity show that it is essential to our humanity, and that we are now genetically 'hardwired for music'.

This rings true with me – far more than the alternative stance from evolutionary psychologist Steven Pinker, who proposes that music is an accident with no useful function for our species (he calls it "auditory cheesecake"). This flies in the face of music's universality. If it were truly useless it would surely have withered and been discarded thousands of years ago. Instead, in the words of leading music psychologist Donald Hodges: "Musicality is at the core of what it means to be human. For, to be human is to be musical and to be musical is to be human."[26]

Just as humans share so many features (including the vast majority

---

* Sound itself has been developed as a weapon by the US Army and Navy, who have experimented with both ultrasonic sound (using beams of sound at high energy levels either to carry unpleasant noise that disables the enemy or even to cause physical damage on their own) and infrasonic sound (which can create sickness and even instant bowel evacuation). There is a whole book on the subject called Sonic Warfare – Sound, Affect and the Ecology of Fear by Steve Goodman if you want to know more.

of our genetic code) with higher apes, language and music have more similarities than differences. Both have structure; both have individual events that combine to create complex forms (language's syllables - words - phrases – sentences – chapters - books compared to music's notes - chords – phrases - melodies – songs – albums, as a rough example); both have syntax; and both can communicate meaning and emotional states. There are significant areas of overlap, such as poetry, where cadence and rhythm are so vital, and chant, especially mantric chant, where the words and the music fuse and it's hard to know which is predominant.

The work of Professor Michael Tomasello[27] on human cognition indicates that we have to learn a range of socio-cognitive skills in order to use language – while our musicality is there from the beginning. In other words, we are born musical but we have to learn language.

It's worth reflecting on the spiritual dimension of music here. We have discussed the universality of vibration – the fact that we humans, like everything else in the universe, are composed of essential vibrations. The principles of entrainment and resonance tell us that one vibration can affect another, so it seems eminently reasonable to speculate that sound vibrations will affect the vibrations within us, changing the state of our component matter in ways we don't yet understand. Every advance in quantum physics seems to move its frontiers further into contact with metaphysical concepts, and this kind of postulation is far less bizarre than, say, the idea that two particles, once they have connected, can communicate changes in their state instantaneously across the vastest distances, or the notion that every experiment's outcome is affected by the very presence of an observer.

A religious perspective would be that music comes from and connects us to God; certainly, music has been used in every religious practice as a channel to the divine, and a potent portal to states of spiritual connection and even ecstasy.

A metaphysical version would be that human music is our natural response to the music (vibration) all around us – what the Sufis call 'the music of the spheres'. Perhaps we perceive this at a non-conscious level, and our human music is a necessary response to affirm our place in the universe.

Wherever music comes from, it exists everywhere there are human beings, and we are now studying it in earnest. In the last century the formal study of the role and effects of music commenced with the work of Hermann von Helmholtz, and the resulting academic fields (psychoacoustics, music psychology, ethnomusicology, biomusicology and more) have exploded in the last forty years. There are now legions of academics studying music from every angle. In Hodges' essential *Handbook of Music Psychology* there is a listing of major music psychology texts and another listing of books in every contiguous field from music and chaos theory to plant music.[28] The rate of publication has increased rapidly in the last four decades.

This level of study is long overdue, because music is entwined through every aspect of human activity. It is found in homes, schools, community events, sport, religion, celebrations, politics, the military, healthcare, physical exercise, marketing, cars, planes, public spaces, commercial spaces, personal stereos, films, concerts, TV – and music is in its own right a huge business, although, as has been well documented, one in great flux. In January 2007, Reuters reported that global online music sales had doubled to $2 billion, 10 per cent of total sales – but this increase had not been enough to stop total industry sales from falling by 3 per cent compared to the previous year. Nevertheless, at $20 billion revenues, the industry is still a significant element in the world economy.

All this money comes from (and stimulates) an ever-increasing quantity of recorded music. I fondly remember a time when, as a music fan, it was really possible to feel that one knew what was going on: the week's important releases could be counted in the dozens. By the 1990s, the position had become very different. University of California's Berkeley College estimated that in 1999 the global production of unique music CDs was around 90,000 units. At 45 minutes per CD, it would take about eight years to play that one year's output back to back. Someone listening eight hours a day, five days a week would take over six months to get through a single week's output. Berkeley further estimated that the world's stock of CD titles was around 1.4 million – which must have grown to well over 2 million by now. We long ago passed the point at which we could keep up with the quantity of music being released.

So music is everywhere, its flow has become a torrent and there are hundreds of people analysing it – but none of the research that I have seen gets anywhere near unlocking the black box. We still have no real idea how music works. This is probably because it simply contains too many parts: melody, harmony, timbres and instrumentations, voice and words, tempo and rhythm, style and associations – all these things affect us, as we'll see when we go into the SoundFlow™ model in detail. When they are all working and interacting at the same time, it is impossible to separate out the individual strands of the spaghetti. All we can really do is eat it and see what happens.

There is plenty of research of that kind, so we certainly do have some ideas about the effects of music on people in various situations. As the world leader in piping music into commercial spaces, the Muzak Corporation has a vested interest in proving that music can create business benefits, and it has done a lot of research to prove it. Hodges and Haack[29] list the following benefits claimed by Muzak to arise from using its service:

- A 29 per cent decrease in nonessential conversation or activities among telephone company employees

- A 32 per cent decrease in lateness and absenteeism among the employees of a giant corporation

- A 39 per cent decrease in errors in the accounts payable section of a business office

- An 8 per cent increase in productivity, even after a bonus system had been installed, in a publishing company

- A 19 per cent increase in key punch productivity at an electric utility company with a corresponding decrease of 32 per cent in errors

- A 53 per cent decrease in airline agent turnover

- An $8.4 million increase over expectations in bank earnings

- A 25.5 per cent better accuracy rate in editing

- A 25 per cent increase in enjoyment of the workplace

- A 16 per cent increase in problem solving abilities.

As Hodges and Haack conclude: "If these figures are the result of rigorous, controlled experimentation, as Muzak claims they are, they give a clear indication of the powerful influence music can have on working behaviours."

There is now a large amount of research from independent, academic sources about the effects of music on people's behaviour in shops, malls and restaurants. The top-line summary is that music does significantly affect people: it can speed them up or slow them down, and it can change their mood. As a result they behave differently – and, most importantly for the retailers in question, it affects the amount of time and money they spend in the establishment.

Sadly, few retailers have taken the time to understand the research properly. In fairness there are often countervailing influences to consider; most of the academic studies aim to measure the effects of only one variable, so in order to predict holistic effects one has to combine these individual influences (and of course cross-modal effects where other senses than hearing are being affected). This is more art than science, which leaves the field open for the smooth selling of music as the one-size-fits-all solution for every commercial space by the music industry and its representatives in the various licensing bodies. Of *course* they want to veneer the whole world with music: the industry is declining (down 7 per cent overall in 2009), and the performance rights market is one of only three growth areas left to cling on to (along with live music and songwriters' music copyrights, which includes the lively new band/brand space where artists are being sponsored by the new patrons of the 21st century – brands).[30]

I love music, and that's exactly why I hate to see it being turned into some sort of aural whitewash. Most music wasn't made to be background sound; it was made by people who care, and who want their work to be listened to, which is why very often it doesn't work to try and ignore it. In sound, intention is always important. That's why music is often not fit for purpose as a background sound.

We'll be looking at the effects of music on people in more detail in the SoundFlow™ section a little later, and adding some practical experience and specific reflections when we discuss sound in all types of space in Part 3.

Natural sound

Wind ●

We hear wind in leaves, in grass, over rock, moving sand and dirt,

---

and against the flaps of our ears. Its dynamic range is huge, from the susurrance of gentle zephyr that offers a moment's relief from the heat of a desert day to the deafening roar of the strongest winds on Earth during the dark winters of the Antarctic. Its sounds subtly define our natural environment, as for example in the difference between the percussion of fleshy, mid-Spring leaves, dry, brittle late-Summer leaves and mid-Winter twigs. We know and respond to these tiny signals instinctively; they help give us our bearings every day. I don't believe anyone has ever done any experiments on the importance of these tiny, myriad aural data flows derived from the movement of the air around us, but I would expect that substituting an inappropriate signal (for example the sound of wind in bare twigs on a summer's day) would create a profound feeling of unease.

Water* ●

Water's main songs are the sounds of rainfall, streams and rivers and of course the sea. As with wind, its range is enormous, from the gentlest burbling of a tiny brook to the overwhelming all-frequency bombardment of a mighty waterfall. Water is life-giving, the essence of our survival; we find its gentler sounds soothing and restful, which is why fountains have always been so popular, particularly in hot and dry places. Many people think fountains are created to look attractive, and certainly they have been raised to high visual art by the likes of Bernini, but their first function has always been to bring the sound of water (the other essential sound of life, alongside breath) to a place without it.

Birdsong ●

Birds, and in particular songbirds, densely inhabit the same regions we do: the temperate and tropical regions. Nobody knows why they sing (notwithstanding the theories you may have heard about territory and mate selection), or why some birds sing exquisitely beautiful songs and other just croak or squawk. Birdsong becomes more and more amazing as you study it: slow down the lightning-fast song of a thrush or of the diva of songbirds, the lyrebird, and you find complex, repeating structures that combine rhythms that would challenge most master

---

* ● The website contains examples of types of water sounds, and also birdsong in various locations, including a section of slowed-down song.

percussionists with pitch sequences and modulations that use more notes, subtler relationships and levels of vocal gymnastics way beyond any human. It sounds like virtuoso jazz played at breakneck speed – and then you remember that this is slowed down to one quarter of the original delivery pace. David Rothenberg's excellent book *Why Birds Sing*[31] goes into this topic in detail, and is highly recommended, though he doesn't consider the psychoacoustic question we need to consider here: what does birdsong do to human beings?

At the most basic level, birdsong tells us that we are safe. We have learned over countless millennia to use the ceaseless diurnal vigilance of the birds, turning them into unpaid guards by virtue of their practice of changing their song, or most often falling silent, if danger approaches. When the birds are singing, all is well. It's when they stop singing that we need to be on alert. I have no doubt that a sudden cessation of birdsong will still create a release of cortisol and adrenaline, the fight/flight hormones, in a modern human being.

Birdsong is also nature's alarm clock. When the birds start singing, it's generally time to get up, so we associate birdsong with being awake and alert. Thus playing birdsong tends to make people feel cognitively alert.

The third effect of birdsong is to connect us with the world. There may be some element of feeling not alone in this, of being in the company of other living things that are no threat to us. Birdsong seems to affirm life and the joy of it (and there is good reason to believe that this is actually why birds are singing for much of the time). I have met only a handful of people who dislike birdsong; most people enjoy it and find it beautiful. It seems natural for us to take aesthetic pleasure in one of the planet's signature sounds – one that has been there far longer than we have, according to current theories. Our developed appreciation of birdsong may be there because listening to it is a significant physical manifestation of our connection with nature – a connection that modern living has severed for many millions of people.

Whatever the reasons, birdsong is enduringly popular. Musicians have always been fascinated by it: Mozart kept a trained starling to listen to, and Messiaen attempted to recreate birdsong in his later music, though most birdsong is impossible to transcribe. It's not only trained professionals who appreciate the uplifting nature of birdsong: in the UK recently, a British Trust for Ornithology CD of nightingale song rushed off the shelves as fast as they could be restocked, as did the British Library's *Dawn Chorus* CD. Maybe this is because, as a study by Reading University found, encounters with the natural world boost mental health by giving 'a sense of coherence.'[32]

For these and other reasons explored in the SoundFlow™ section (such as the likelihood that high frequencies charge our neural system up, refreshing us) we have used birdsong as an important element of soundscapes that we've created for several clients, including an airport terminal soundscape for BAA and the default soundscape for BOX, London's high-level specialist consultancy workspace. *●

WWB

As we have seen, the combined soundscape of wind, water and birds (WWB) is primarily stochastic and, after hundreds of thousands of years' practice, we effortlessly apply differencing to move it to the background. In my opinion we have also developed a symbiotic relationship with this type of sound. It makes us feel comfortable because it has always been there. Most of the time it was the *only* sound: the odd war would make a lot of noise, and there were loud local events like blacksmith's hammers or church bells, but all these were noticeable mainly because they were relatively rare compared to WWB. Anywhere you found humans on the planet, the soundscape was dominated by one or more of the components of WWB.

It's only in the last 250 years, since the Industrial Revolution, that human beings have started to live in places where none of the three primary stochastic sounds exist. From this time onwards the soundscapes of our cities changed dramatically – and they are certainly not stochastic. They are composed of much starker, more noticeable sounds, like road vehicle engines, tyres and horns, trains, planes, a plethora of varied warning tones, and other people's conversation. Most of these are above our differencing threshold, and none of them (save generalised traffic noise) is stochastic in the sense we are using here. Not many street soundscapes merge pleasingly into a strangely comforting wash!

I believe the removal of WWB and its replacement with non-stochastic

---

* ● There are examples of both these soundscapes on the website, along with a loopable section of birdsong for you to use at home or at work for rest or for stress-free working.

urban soundscapes have created two results. The first is stress. Instead of using the effortless resources we've developed over hundreds of millennia to move WWB to unconscious listening, our sound processing system is having to work overtime to suppress a barrage of noise all day. This is hard work, and not surprisingly it's tiring. Many people have to work in places where the ambient noise level is well over 80 dB, and there is plenty of research to show that their health suffers. (For more on this, see the section on Staff Spaces in Part 3.)

The second result, I believe, is that we are pining for WWB. This was beautifully illustrated when I was buying a sofa for our office some time ago in a small furniture shop in London. The charming, knowledgeable and elderly salesman and I were trying to negotiate terms over the top of BBC Radio 2, which was being piped loudly to all parts of the store. As I often do, I asked if the music could be turned down, a request to which he responded with enthusiasm. When we could hear ourselves think, I asked him why they had the music at all. "If it was up to me we wouldn't," he said. "But people like some noise these days."

I suggest that the removal of WWB has left us with a vague feeling of loss: we know we ought to be listening to something, but we don't know what it is – so we put on music, or the radio, or the television. As noted above, birdsong is an excellent alternative: at BOX, people who haven't even noticed the birdsong remark on how fresh they feel at the end of a hard full-day workshop.

A respect for, and understanding of, the significance of WWB for human beings is an essential tool to take into the business of designing soundscapes for the spaces we inhabit today.

Noise

Noise is becoming a major issue in the modern world, primarily because it has been growing continuously for many years. Perhaps it has reached a threshold at last, a level beyond which people are not willing to allow it to go on increasing. Modern city soundscapes can be as loud as 90 dB, and are generally over 80 dB, and they have been getting louder every year. Estimates of how much louder vary, but Murray Schafer is usually a reliable guide in these matters and he reckons the rate of growth is half a decibel a year.[33] This means that cities are twice as loud as they were twenty years ago.

What do we actually mean when we use the word 'noise'? It's not a concept that is unique to the world of sound: scientists and engineers are familiar with noise in all sorts of systems and environments, from computers and electrical circuits to cosmology. One simple definition from the world of engineering starts with a simple duality between signal and noise. Signal is all the information we want. Noise is a residual in this view: everything that is not signal is by definition noise. Another definition is that noise is simply unwanted signal.

This is all somewhat subjective. In his fascinating book *Noise*[34], Bart Kosko emphasises that one person's noise is often another person's signal, as in the increasingly common urban phenomena of the noisy neighbour or the train carriage mobile phone conversation. To the average hardcore punk, Beethoven is unpalatable noise; to a devoted classical music buff, the Sex Pistols may be noise incarnate.

The purest form of noise is called white noise. Most people have heard of it, but not many know what it is. In fact, pure white noise cannot exist because its definition is a sound with equal power across an infinite range of frequencies. That would require an infinite amount of power, which is of course impossible. In practice, white noise is sound with equal power across all the audible frequencies. White noise is perfect noise: it is the same everywhere, so it has no signal (or it is all equally signal). We perceive it as predominantly hissy, but this is just because of the uneven sensitivity of our hearing: we are more receptive to higher frequencies.

Pink noise compensates to some degree for that unevenness by increasing the power logarithmically as the frequency increases, applying equal power in bands that are proportionally wide – for example the same amount of power from 20 Hz to 40 Hz as from 2 kHz to 4 kHz. In consequence its actual power declines in a straight line as frequency rises, roughly matching the increasing sensitivity of our hearing. We hear it as a more broadband noise. It is often described as white noise adjusted to sound flat to humans, but this is in fact not the case. It sounds flatter than white noise, but for a truly flat sound we have to turn to one of the many other colours of noise: grey noise.

Grey noise is white noise with an inverted A-weighting. Although we have noted that A-weighting is not a flawless map of human hearing, the effect of boosting all the frequencies we don't hear so well and reducing the ones we are most sensitive to is undeniably to produce a much flatter quality of noise. *●

There are even more colours of noise, including brown, blue, orange, green and even black. None of them are of interest for our purpose so we'll leave them in the textbooks.

In some situations noise can actually be beneficial: for example, stochastic resonance is the principle of adding some noise to a non-linear system and thus improving its performance. Masking sound is a practical example of this process.

However, most often noise is a problem. Noise is becoming omnipresent and as we've noted is ever-increasing in urban areas. It is now a major concern for governments because it has serious social and economic consequences. Noise limits our channel capacity and is fragmenting our society. As Kosko notes: "A high level of background noise partitions space into many small acoustic arenas."[35] These are typically created by the simple expedient of a pair of headphones and a personal stereo.

The World Health Organisation has published guideline maxima for daytime and nighttime noise exposure (55 dB and 45 dB respectively, both A-weighted averages over the relevant period). It is clear that many people's health is being damaged by exposure to levels that exceed these maxima. When Building Research Establishment (BRE) carried out the UK National Noise Incidence Study for the Government in 2000, it reported that "the majority of the UK population are still exposed to noise levels exceeding [the] WHO guidelines."

Across Europe the same picture is seen: the WHO study *Community Noise* by Birgitta Berglund and Thomas Lindvall Stockholm, Sweden, warned in 1995:

"Almost 25% of the European population is exposed, in one way or another, to transportation noise over 65 dBA (an average energy equivalent to continuous A-weighted sound pressure level over 24 hours) (Lambert & Vallet, 1994). This figure is not the same all over Europe. In some countries more than half of the population is exposed, in others less than 10%. When one realizes that at 65 dBA sound pressure level,

---

* ●You can hear samples of white, pink and grey noise on the website.

sleeping becomes seriously disturbed and most people become annoyed, it is clear that community noise is a genuine environmental health problem."The *English Housing Survey 2008-09 Household Report* found that: "Road traffic was the most common cause of noise problems and was reported by 4.6 million households (22% of households). This was followed by 2.6 million households (12%) reporting noise from other neighbours in the street to be a problem, and 2.4 million (11%) reporting noise from immediate neighbours or common areas of flats to be an issue." (One source only was allowed in the questionnaire.) The US is no better. According to a 1999 US Census report, Americans named noise as the number one problem in neighborhoods. Of 102.8 million reporting households, 11.6 million stated that street or traffic noise was bothersome, and 4.5 million said it was so bad that they wanted to move. For the category 'other bothersome conditions,' 2.7 million named noise. Additionally, 5.3 million said that they were bothered by building neighbour noise. More Americans are bothered by noise than by crime, odours, and other problems listed under 'other bothersome conditions.'

When the European Union launched the current Europe-wide programme of noise-mapping to help inform strategies for noise reduction, it said: "Environmental noise, as emitted by transport, industry and recreation, is reducing the health and the quality of life of at least 25 per cent of the European Union's population."

The effects of noise intruding into people's home soundscapes are well charted. Loss of sleep is the most significant, with all its associated symptoms. But there are also irritability, reduced sociability, reduced communication, and possibly increased blood pressure, adverse blood chemistry and other medical effects.[36]

Noise interference with verbal communication (for example aircraft noise interrupting conversation) has major effects without needing to deprive people of sleep. In their 1995 WHO report, Birgitta Berglund and Thomas Lindvall said: "Noise interference with speech discrimination results in a great proportion of person disabilities and handicaps such as problems with concentration, fatigue, uncertainty and lack of self-confidence, irritation, misunderstandings, decreased working capacity, problems in human relations, and a number of reactions to stress."

As discussed in the section on office sound later in this book, noise also has an adverse effect on social behaviour. The US report Noise and its effects (Administrative Conference of the United States, Alice Suter, 1991) says: "Even moderate noise levels can increase anxiety, decrease the incidence of helping behaviour, and increase the risk of hostile behaviour in experimental subjects. These effects may, to some extent, help explain the 'dehumanization' of today's urban environment."

The cost of invasive noise to society and to business is staggering. The official EU estimate: "Present economic estimates of the annual damage in the EU due to environmental noise range from €13 billion to €38 billion. Elements that contribute are a reduction of housing prices, medical costs, reduced possibilities of land use and cost of lost labour days. In spite of some uncertainties it seems certain that the damage concerns tens of billions of euro per year."

A good proportion of the enormous cost of noise is undoubtedly falling on business in the form of lost productivity, paid sick leave, absenteeism, antisocial or unproductive behaviour and the cost of mistakes made by noise-affected employees. It is well documented that people who are short of sleep have slower reaction times and make more mistakes; the EU estimate above, vast though it is, does not include any of this huge extra cost.

This is a classic case of the competitive economy failing to work because the indirect costs of an activity are disassociated from the action itself. Economists have struggled for years with the problem of what they call 'negative externalities', for example a loud party where everyone present is having a great time but the neighbourhood is suffering the fallout in the form of lost sleep, irritation and frustration. We need to create feedback loops that promote self-regulation of these social evils, in much the same way that legal and taxation regimes have been developed to combat chemical and other forms of environmental pollution. If companies were required to account publicly for the billions in noise costs caused by their core activities like transport, construction and manufacturing, we would certainly see some changes.

As I've already suggested, most heavy diesel vehicles are inordinately noisy simply because nobody has ever asked the manufacturers to make them quieter. Like air pollution in decades past, it has not been a feature much considered when choosing a fleet of trucks or buses. We have seen a quantum change in carbon emissions since specific laws and taxes were introduced to incentivise people to make and operate cleaner vehicles.

Now we need the same techniques to incentivise them to make and run quieter vehicles.

Of course it's not just corporate noise that invades the domestic soundscape. Commenting on the EU noise maps, *Newsweek* noted: "A single noisy motor scooter driving through Paris in the middle of the night can wake up as many as 200,000 people."[37] However, aside from noisy neighbours, the top noise nuisances – planes, trains, road traffic, construction and heavy industry – are predominantly generated by organisations, not by individuals; it seems clear that organisations are responsible for most of the noise invading our homes, as well as suffering most of the hitherto hidden cost. I hope this book will help to link the activity with the cost, and to introduce the idea that less social noise will lead to higher profits, as well-rested and relaxed employees will do better work faster and with less mistakes.

The main practical problem until recently has been the lack of ways to quantify ambient noise and its effects. Europe's noise maps, which will eventually cover the whole of the EU and, one hopes, spur the rest of the world to follow suit, are proving invaluable in identifying the most blighted areas so that we can at last get to grips with the worst of ambient noise. They are also an important manifestation of society's communal desire to push back, to find more and better mechanisms to quantify this social bad and to levy its cost against those creating it so that they have an incentive to act in other ways.

London's ambient noise strategy was another step forward. The 295-page plan, entitled *Sounder City* and published by the Mayor's office in March 2004, was the first such strategy published by a city authority and it informed policy on public transport, which is a major noise generator. New, quieter buses have been on trial in London, including hydrogen fuel cell and hybrid-electric vehicles, and there are more initiatives to come. The strategy covers every aspect of noise that comes under the control or influence of the city authority, including aircraft, industry, police vehicles, refuse collection, commercial vehicles, traffic and local noise legislation. In time perhaps every major city will have such a plan.

Ultimately, though, repelling the sonic invaders of our homes will be the job of governments, using the twin tools of legislation and taxation. The problem with large social evils like noise pollution is that there is no point in one organisation changing its ways if all the others are going to carry on as before: the improving organisation would simply be disadvantaged (except in special cases, for example if its noise were affecting mainly its own workers). And so the situation persists, worsens even, until the rules of the game are changed for everyone.

Governments will act only when they sense enough desire for action among their voting populations. This means it's in all of our hands. I feel optimistic because noise is climbing the political agenda year by year, and because we are developing the tools that will show us the damage that's really being done. With this knowledge, and the practical experience of noise blighting so many homes, we can hope that the public will focus on the problem, the politicians will act, and that ambient noise levels will peak in the coming few years and then start reducing in the second decade of this century.

Noise at work

Inside working spaces, noise can adversely affect productivity, morale, motivation, teamworking and health. In high-noise occupations, the effects are known to include headaches, fatigue, gastric problems such as stomach ulcers, increased blood pressure, stress, and excessive exposure to the fight/flight hormones adrenaline and cortisol. In offices, research shows that noise leads to reduced productivity, stress, unwillingness to help and communicate and other undesirable effects. How strong these effects are depends largely on the three Cs: control, contrast and conversation.

Control

The unstoppable rise of the personal stereo stems at least in part from the desire of every human being to retain control over his or her personal space, including what they hear. In essence, the iPod is a defence against intrusion. (We look at this in more detail in section on personal soundscapes.) The same process operates at work: if people cannot exercise some control over the noise around them, it becomes an intrusion and creates stress. Control could be turning the volume down, switching a device off, asking for quiet or moving to a quiet space. Without the ability to control noise, people become upset, stressed, negative and less productive.

Contrast

There is evidence that constant noise is eventually habituated: in other words, if it's constantly loud, people can adapt. This is vitally related to contrast, or variability. If the noise is unchanging or highly repetitive (like a loud machine) then the brain can adapt after a time. If it varies, and particularly if it stops and then restarts, habituation is destroyed. Also, people who are not used to noise find it more upsetting than people who have already habituated. Finally, there is no research yet on the long-term effects of acclimatised exposure to constant loud noise, but it is reasonable to assume that this must be fatiguing and unhealthy over time.

Conversation

We saw in our review of listening why conversation is the most distracting noise of all. When you can hear what someone else is saying, you cannot stop part of your brain from paying attention – and your performance suffers. It is not just quantity of noise that matters: if the noise is mainly intelligible conversation, it can be relatively quiet and still cause great disruption. We'll review the effects of noise in working spaces in detail in the SoundFlow™ section and also in the relevant parts of Part 3.

The sound of silence

Most people would say that silence is the opposite of sound, but Dame Evelyn Glennie takes a different view and I agree with her: silence is a sound, as well as being a context for all other sounds. To understand this it is necessary to experience silence fully. This takes some effort now, but it is worth it. I cherish memories of moments spent in complete silence. On retreat at Worth Abbey in Kent, UK, it is possible to sit alone late at night in the huge circular modern church, much of it built underground. There is just one spotlight on the central alter. After the echoes of one's entrance and movement die away (the reverberation time is impressive) the silence settles like a thick garment, pressing in, gently insistent but never oppressive. The sense of time passing fades and the moment stretches into eternity. This kind of silence is an experience to be embraced, essential for a real understanding of listening. Finding it is not easy as buildings like Worth Church are rare and other truly silent places, like deep caves and remote wildernesses, are often hard to get to or occupied by other people who don't have the same objective and cheerfully make noise to avoid precisely the experience I am describing. But I commend it to you as something to do at least once a year – like cleansing your palette or detoxifying your body, it enriches your appreciation of the usual, and in some way resets you and helps you to cope better with modern living.

I did just that a while ago on a short holiday in mountainous Northern Italy, where my wife is from. Doing what I do, I naturally listen to every place I visit and on this trip three experiences made me rediscover the value of silence.

First was a visit to *Isola S. Giulio* in the middle of beautiful Lake Orta, near Milan. This small island houses a basilica and a convent for a community of nuns of a silent order, which is why it's known as 'the island of silence'. Encircling the island is a single footpath: *La Via del Silenzio*. Visitors are encouraged to walk the path in silent reflection, and every hundred metres or so there is a board showing one meditation on silence for the way out, and on the other side one for the way back. I was struck by these meditations because they are so universal. There is no hint of Catholic dogma; rather, they resonate with the deep wisdom mined by every spiritual path that has discovered the power of silence – which is most of them. Here are the meditations:

- In the silence you accept and understand
- In the silence you receive all
- Silence is the language of love
- Silence is the peace of oneself
- Silence is music and harmony
- Silence is truth and prayer
- In the silence you meet the Master
- In the silence you breathe God
- Walls are in the mind
- The moment is present, here and now
- Leave yourself and what is yours

Walking the path and internalising these reflections created a sense of deep peace and wellbeing, and of being fully present in the moment - which is probably saying the same thing in two ways.

Second by dramatic contrast was Milan's railway station. This is a monumental building from Mussolini's time, built on massive scale and with the acoustics of a cathedral. Sadly its grandeur is being eroded by the recent installation of many plasma screens showing a looped couple of minutes of advertising – with sound played through the station PA system. At first I thought they were playing opera, until the fragment repeated again and again as a small part of the loop, advertising a mobile phone service. Opera in that space would have been interesting, pleasing and, with La Scala close by, very appropriate. The looped advertising sound felt intrusive, overbearing, irritating and even profane in that grand building, adding a gratuitous extra level of noise to the existing reverberating cacophony of train engines, footfall, voices and sundry machinery. (Incidentally, all the subway stations have two large projectors on each platform, again with sound booming out of them. Thank goodness that in the London Underground the projectors now being installed are silent.) Milan is a very worrying example of what could be the future in all public spaces if we're not careful. Never did silence seem more valuable than in this awful noise.

The third experience was high in the awe-inspiring Dolomites, which to me are the most beautiful mountains on the planet. We trekked for three days, staying at rifugi up to 2,500 metres above sea level. The air was like crystal, the views were overwhelming and from time to time we heard the silence of the mountains. In my experience, the deep silence of nature is to be found only in high mountains or in deserts (hot or cold), because in these places there are almost no birds or insects. When the wind dropped and in between the infrequent high-altitude planes, the Dolomites offered us that rare experience. The deep silence of nature is rich and pure: it is the essential context for all other sound, just as a dress in black (the absence of all colour) is the context for what it contains. This silence is the sound between all sounds. Immersed in it, one can start to sense connection and resonance with all of nature.

There are unquestionably different kinds of silence. At the extreme is an anechoic chamber. With no sound source and zero reverberation, this is the purest silence humans can achieve (because we can't survive in a vacuum, the ultimate silence). However, after a short time in such intense silence one starts to hear internal sounds: blood pumping, lungs and other organs moving, tinnitus in the ears. In the end, this overbearing artificial silence does not offer us the experience of silence at all.

In a truly silent building such Worth Church, overtones define the shape of the space. With eyes closed and without any sound, you can sense you are in a huge room. Indoor silence like this is rare and to be cherished, and is wonderful for meditation, prayer, contemplation, or even working. It has an entirely different quality to the silence of the mountains, resonating with all that is best about humanity rather than a deeper connection with nature.

The silence of nature is to me the finest of all, because in it we sense our connection with everything. However, it's becoming a precious commodity. If silence was golden in the 1960s, it's a rare and precious diamond now. There are few remaining wildernesses that offer more than a short burst of true silence. Nature recordist Bernard Krause claims there is now almost no place on Earth – including the North Pole, Antarctica and the dense forests of Indonesia and the Amazon – that is free of aircraft overflights, the buzz of chain saws or other human clatter. Krause remembers when it took 20 hours to get 15 minutes of usable recorded material. "Now it takes 200 hours," he says.[38]

There is a third kind of more accessible silence, simply defined by lack of proximate speech and machinery, especially cars, planes and trains. This is the silence one can experience at Orta: the soundscape is in fact quite rich, with lapping waves, birds, wind, and even distant human sound such as boats and high planes. It's not total silence, but in this quietness there is still peace, as we found when walking the Way of Silence.

In cities, silence is something that most people actively avoid. Their first reaction on walking into a silent room is to turn something on – radio, TV, stereo, anything to stop the silence. They have become so used to urban noise that they feel uncomfortable without it. I think urban living has created an addiction to noise as a means of avoiding being fully present.

Silence is a medium for growing human consciousness, an invitation to be fully present, and a doorway to a sense of connection with the universe, or God if you prefer. How sad that we have made it an endangered habitat – and that this process is accelerating. Will we in future trek across mountains wearing our iPods? Have we altogether lost the desire to be present, connected and conscious? Or can we preserve the silent places and benefit from them in the ways of our ancestors?

2.3 The Golden Rules of sound

Wherever a soundscape is designed and deployed, it should always be created in accordance with four Golden Rules of sound.

1 Make it optional

Among the tiny proportion of websites that were early adopters of sound, there was a practice of forcing sound on users whether they liked it or not – and even continuing to play it after they had hurried on to the next site. Thankfully it's getting rarer to encounter this nonsense; most sites have a very visible button offering a toggle between sound on and sound off. This is vital, and it's a principle that should be adopted wherever possible in the real world too.

The backlash against music in public places (more about this in the section on sound in shops) is fuelled by the resentment that arises from being given no choice. We saw in the section on noise that the people's irritation increases dramatically when they have no control over the sound source. It follows that we must aim to give people a choice about the sound we inflict on them.

Obviously this is difficult to do in a physical space – though not impossible. Zones with different sounds are one practical solution, as educational establishments with silent reading rooms have long understood.

If we can't offer truly optional sound, the next best thing is to target our sound as carefully as possible, so that we upset the smallest number of people. For spaces with a very tight demographic and psychographic user profile, this is not too difficult. Some shops, bars, clubs and restaurants know exactly who their customers are and what they like; in many cases the sound (usually music) acts as a filter, attracting the 'right' people and warning the 'wrong' ones to go elsewhere because this is not for them.

This approach can work in more generalist spaces if music is used as part of an overall zoning policy. For example in a large mall there might be zones for younger and older customers, and music could be a form of signposting to help nudge people in the right direction – maybe club music in the former section and jazz standards in the latter.

The problems arise for generalist spaces that can't or won't operate this kind of zoning. As we have seen, one person's signal is another person's noise, and nowhere is this more true than with music in public. Whatever you play in a mass-market space, you will upset someone. I strongly suggest two actions. First, err on the side of caution: it's better to inject no sound that the wrong sound. There is nothing at all wrong with the sound of people shopping! Second, research carefully before you deploy. Do not let the smooth patter of a music-streaming company persuade you that your customers will naturally love smooth jazz and R&B classics, because they just might loathe them. Use focus groups to ascertain attitudes, and most important create pilot sites where you run proper quantitative tests that measure the effect of the sound on people's behaviour. It's what they do that matters, not what they say.

2 Make it appropriate

We've looked at how to map out the best applications of sound for a brand, and how to define what that sound should be. When you've done that piece of work you will have no trouble in making sure that all the sound you inject into your spaces resonates with your own organisation, brand, products, values, image, practices and so on. This is the first test of appropriateness: is this sound right for us?

The second, of course, is: is this sound right for its context? This is where we explore all the four modifiers in the SoundFlow™ model (described in detail in section 2.5), taking care to ensure that whatever we design fits with the space's function, environment, people and values.

3 Make it valuable

There are far too many shops playing music because they do it next door. I suspect that the world would sound rather different if they all asked the question: what is the value of this to our customers?

Sound can be hugely valuable. It can warn us of danger (smoke alarms); it can inform us of events or of opportunities (radio news; in-store announcements of special offers); it can reduce the boredom of mundane tasks (music in factories); it can entertain, move and inspire us (music); it can guide us (zoning; travel announcements); most of all, it's our primary connection with other humans (conversation). When designing a soundscape, always ask how sound can add value for the customer. If you can't answer that question, silence is golden.

4 Test it and test it again

I have said several times that when it comes to sound it's what people do that matters, not what they say. This is particularly true when the sound in question is music, because everybody has an opinion about music. If you send researchers out into a space to ask people what they think of the background music, the typical conversation will go like this:

> Researcher:"Hi, can you spare a minute? We're asking people what they think of the music in this store."

> Customer: "Music? I didn't... oh yes, now I notice it." (Eyes glaze over as opinion is formed.) "Well, I personally don't like [style of music currently playihg]; it would be much nicer if they played [favourite style of music]."

I've had that conversation many times. Before the question was asked, the music existed only at the unconscious level. It was still having an effect. Bringing it to the conscious level by asking about it changes the relationship completely, and what you get is preconceived ideas, not a valid reading of how the music was for that person before you interrogated them.

I have found that only two kinds of research into people's opinions are useful. Qualitatively, it's interesting to run focus groups of customers (or, for larger audiences, customer segments) to understand what sounds they like – not just music – and what they dislike. Auditory 'mood boards' and specific sounds and music tracks can be used as stimulus material. Quantitatively, it is then useful to research the right demographic groups in larger numbers to get corroboration of our qualitative selections – but this should not be done in relation to the experience of the sound itself. All of this research is useful because it tells us what not to include in the soundscape.

Designing effective research

Our standard methodology when testing the effect of a soundscape on behaviour – typically in retail – is to create a symmetrical data set. The number of locations will be two, four, or any multiple of four. If we are testing one soundscape, we set it up to run every other day. If we are testing two alternative soundscapes, we set them up to run on alternate days. Either way, we have two states to compare – either soundscape v nothing or soundscape 1 v soundscape 2.

On the first day, we start half the sample in one state ('on') and half in the other ('off'), and then we run the test for a period of four weeks (or any multiple of four weeks). Having half the sites on and half off on any given day means that any external shocks to the system are normalised as they affect both on and off states simultaneously and equally. Any daily trading patterns are normalised because we have an equal number of Mondays (for example) in both states.

What we are measuring is usually retail sales, divided, if possible, by number of customers to produce sales per head. If the customer numbers are not known, we simply use total sales. At the end of the test we compare the average of all the data points in each state to measure the effect of the new soundscape.

Of course sales are both the retail client's prime objective and also a proxy for several other important factors that form a chain between the soundscape and the amount going into the tills: customer satisfaction (and its close relative, brand affinity), which tends to be correlated with dwell time, which in turn is correlated with sales in most cases. If we can directly measure dwell time, all the better. However unless there is omnibus methodology in place that we can just make use of, the cost of doing this is usually very high; all retailers measure sales religiously, so this method gives the best bang for very few bucks.

If possible, and always if the space is not a shop, we will look to measure experience more directly by asking questions about feelings of wellbeing; enjoyment of the overall experience; brand affinity if relevant; intention to return, and so on. We never mention the sound. This way we get an unskewed answer and can compare the variables between the two soundscape conditions.

2.4 The sound of the future

Generative sound

One major problem with creating a soundscape from recorded material is boredom, especially for people who spend a lot of time in the space, like shop staff. By definition, recorded sound is static, unchanging and identical every time you hear it. When we look at sound in shops, we'll see some of the effects of playing limited selections of music over and over again. If music is the best soundscape, the playlist must be long enough to avoid boredom, and it must be shuffled often.

For other soundscapes, a more interesting approach is to explore generative sound. I believe that this is the future of soundscapes for commercial and domestic environments. Generative sound is composed and played live by a computer according to rules programmed into it, using either sounds it creates (through a tone generator or an internal soft synthesiser) or recorded sound it has access to through an associated sampler. The rules are usually probability-based – for example, after any note has been played, there may be rules specifying the probabilities of each interval to the next note. This kind of rule can enforce a mode, or create harmonic character, while still allowing chance to create ever-changing musical sequences. Rules can be used to control melody, harmony, volume, velocity, instrumentation and arrangement.

We can create this kind of soundscape today largely thanks to Tim and Pete Cole of SSEYO, whose Koan programme was the pathfinder in this field. Brian Eno had conceived of generative music in the 1980s and was delighted when the Coles approached him with their first pass at Koan in 1994. He worked with them to develop it, and composed a number of generative albums using this software*.

---

* Tim and Pete have since developed Noatikl, a generative VST plug-in for composers and sound designers. There are more complex programs available, from the sonic artist's favourite MAX/MSP to the esoteric SuperCollider, which sonic artist Mileece used to create some beautiful sound for our BOX installation (and also on her fascinating album *Formations*). She's the only musician I know who composes entirely in code. At The Sound Agency we have developed our own generative system called the Ambifier™, which is custom-designed to create ever-changing

At The Sound Agency we specialise in creating generative soundscapes that never repeat, are always changing and always doing new things, though within a clearly defined style. Three examples are available on the book's website: a generative soundscape we created for the new Shetland Museum; what I believe to be the world's first piece of generative music on hold, composed for Chemistry Communications; and the generative soundscape created for Glasgow Airport.●

The Shetland soundscape is designed to give visitors an experience of the auditory richness of the islands. It has three scenes – land, shore and sea – and for each of the scenes the computer has several 'bags' of sounds to pick from, and it picks and plays only one sound per bag at a time. There is no chance of hearing a whale and a tractor at the same time, because each scene's sounds are discrete; also, the bags allow us to stream sounds so that we can create pleasing combinations that don't overtax the listener's bandwidth: for example, human voices are in one bag so the computer can't play several voices at once. The effect is to create an ever-changing montage that gives people a condensed but realistic experience of Shetland's rich soundscape. Over time, individual sounds will be refreshed on the computer's hard disk (just by replacing samples with new ones with the same name) so that the details, as well as the structure, will change.

The Chemistry piece resulted from a careful briefing process where we iterated towards a musical style they feel comfortable with, given the nature and values of their brand. The programming used a combination of samples, MIDI sequences (also probability-driven), soft synthesiser sounds and rules for the arrangement to create a piece of music that never repeats but stays in style.

For BAA we created two generative soundscapes: one for day and one for night. The daytime piece combined generative birdsong (in which we now have great experience) with gentle, ambient music-style tonal elements using long-attack sounds; the night time piece replaced the birds with soothing water sounds (birds don't sing after dark so it would be double schizophonia to create a soundscape of artificial birds at night!) and more human musical sounds, such as gentle guitar and

---

aural wallpaper for commercial spaces, replacing mindless and inappropriate music with something far more beautiful and effective.

piano tones, to give the impression of a person being close by at this lonelier time. There are long sections of both day and night versions on the website. I find them ideal for restful working or just relaxing at home.

I prefer generative sound to music in many settings because it is always fresh and because it creates fewer associations, so we can predict the psychoacoustic effects more easily.

Interactive soundscapes

One step on from generative sound is the interactive soundscape: it not only invents itself on the fly, but also reacts to what happens around it. The stimuli may be anything that can be digitised: sound (picked up simply by placing a microphone in the space); light (using light sensors so that the system changes with ambient light outside the building or the lighting level inside); weather (rainfall or temperature sensors outside); motion, movement or proximity – all these can trigger whatever sound the designer feels is appropriate. The possibilities are limitless, and I find this the most exciting technical development around. I see a future where every significant space has installed a range of interactive soundscapes, selected and configured to match or change our mood as we wish, given all the other factors in our environment.

Where we now habitually turn on the TV or a music system for some sound, in the future we will have many more choices. We will be able to be woken in many more pleasant ways than the harsh bell of an alarm clock; to create morning soundscapes that charge us up and get us ready for our day; to work in soundscapes that are properly designed to support, nourish and refresh us; to come home to soundscapes we've designed to sooth or stimulate as required; and at night we will be able to be lulled to sleep by whatever sound we find most restful, until the system detects the rhythm of our sleeping breath and turns itself off. There are already mobile apps that do parts of this, playing soothing sound as a lullaby and using movement while sleeping as a proxy for brain wave phases, so that they wake users with a soft alarm when the transition from sleep will be easiest, instead of brutally crashing in to deep sleep.

The idea of suitable sound for the time of day is not new: Indian classical music has long had morning, afternoon and evening ragas specifically designed to be effective at that time. What is new is the idea

of the sound altering its own nature according to what's going on –
so for example our loud and lively evening soundscape would hear the
phone ring and duck its own volume right down so we can hear it too
and have our conversation without having to run for the volume control.
We already have this capability in our cars, so why not in our homes?

Sound delivery technology

Alongside poorly chosen content, the biggest problem we come across in
our audits of commercial sound is low quality sound equipment. All too
often the choice of amplifiers, digital signal processing (DSP), cabling
and loudspeakers is delegated to facilities or even IT departments, who
may have little or no interest in sound and treat the task with the same
lack of aesthetics they would apply to buying a server or a power socket.
Please, please remember that the best sound design will be wasted if it
is delivered through poor or inappropriate equipment. It is absolutely
worth showing your customers and staff that you care by investing in
sound systems that deliver quality sound in the right way.

The future of commercial sound is distributed and digital, using
Cat5 cabling to link intelligent nodes, and routing quality sound from
multiple sources to multiple outputs with web-based control, integrating
alarm and announcement systems with soundscapes and playlists and
adapting to changes room noise by using autogain throughout. If you
are planning now, this is the way to go, and companies like Oregon-
based Biamp Systems are the people to talk to. The danger is that the rise
of video communication will relegate sound to an afterthought, which
is why many very expensive meeting rooms are virtually unusable for
conference calling: nobody thought about acoustics or the accurate and
optimal installation of the sound equipment, so calls are destroyed by
room reverberation, inaudible voices or echoes on the line, all of which
can be avoided if sound professionals, not electricians or video experts,
are the people who install the sound equipment.

There are some exciting developments in sound equipment that make
this even more imperative. Even good quality traditional loudspeakers
are far from ideal in many commercial spaces either because they create
hotspots when a more diffuse sound field is needed (sound is often too
loud just under them, but inaudible in between them), or because they

broadcast sound when a tightly controlled pool or beam is required, for example in front of a video screen or a work of art. Now we have technology to address both of these issues.

Since the 1960s people have investigated the concept of using hypersonic sound waves as carriers for other frequencies. This is an interesting concept because the directionality of sound varies with its wavelength. We can get away with one big subwoofer in a home cinema because low frequencies (long wavelengths) are very non-directional: we perceive them as coming from all around, even though they actually have a single source. The higher the frequency, the shorter the wavelength and the more directional the sound; as we move up to ultrasonic frequencies, we start to talk in terms of well-defined sound beams. So far, interesting but useless: we can't hear ultrasound. The breakthrough came through the concept of using ultrasonic waves as carriers for audible sound, just as FM, or frequency modulated, radio waves carry the sound that we hear when we listen to a radio. As the ultrasonic beam passes through air, the signal becomes demodulated and thus audible. The effect is a beam of sound: if you are in the beam you hear the signal right next to your ears because those air molecules are demodulating it: it feels like wearing headphones but without the headphones! If you are not in the beam, you hear nothing. If you bounce such a beam off a wall, the sound appears to come from the wall, not the source.

Until recently the cost of implementation has been high and the quality low. This has changed in the last ten years, and there are two rival systems on the market at the moment that both deliver reasonable quality (for certain applications) at a sensible price.

I first encountered this startling technology when I saw US inventor and entrepreneur Woody Norris demonstrate a hot-off-the-press hypersonic speaker at the TED conference in Monterey in February 2004, and I was delighted to do the same at TED Global in Oxford in 2005. The speaker projected the sound of running water in a beam around the theatre, inaudible to anyone not in its direct path.

Norris's San Diego-based American Technology Corporation claims that it invented this technology – but so does rival Boston-based inventor Joe Pompeii, with his company Holosonics. Both inventors have won awards for their work; both sell working systems with fistfuls of patents; both have impressive client lists. ATC has worked closely with the US military and signed a licensing deal with General Dynamics; Holosonics installed its Audio Spotlight systems in Disney's Epcot centre. Both systems improve in quality and fall in price every year as demand rises and economies of scale start to work.

ATC also developed a more aggressive form of hypersound using high-intensity beams of unpleasant noise from a much larger Long Range Acoustic Device (LRAD); the ear splitting noise beam from this device, which can deliver 95 dB of sound at two thirds of a mile range and up to 140 dB at closer range, was successfully used in defence against an attack by pirates off the Somali coast in November 2005 by the cruise ship *Seabourn Spirit*.[39]

Versions of this device have obvious applications in various emergency and security situations such as a fire warning or crowd control, where they would seem to be preferable to water cannons and rubber bullets. For more aggressive full-scale military use, the sound projected is babies crying (backwards) at 140 dB. ATC's CEO Woody Norris claims, "[For] most people, even if they plug their ears, it will produce the equivalent of an instant migraine. Some people, it will knock them on their knees".[40] Earplugs are no defence because the beam delivers its crippling effects through skull bone conduction even if the ears are covered.*

More constructively, directional sound from hypersonic loudspeakers or from phase-coherent arrays of smaller traditional speakers is proving its social worth in many locations. Museums and art galleries in particular love it, because it is at last possible to deliver audio commentary on a particular work just to the people standing in front of it. Our generative Shetland Museum soundscape is delivered through a hypersonic speaker, and we have also used them in commercial environments such as the beautiful cabinet of curiosities in London's BOX; the sounds there are triggered by a variety of sensors as people view or touch the artefacts, adding related sounds to create a multisensory experience in that spot only.

Future applications mooted by the inventors of this kind of technology include individual sound pools in cars, or in houses – so making possible

---

* We won't enter here into a discussion of the disturbing uses of sound as a weapon, but if you are interested in the topic there is a whole book on it: Sonic Warfare: sound, affect & the ecology of fear by Steve Goodman, MIT Press, 2009.

'his and hers' listening while side by side in bed, or allowing one person to listen to music while another watches TV.

We are some way away from this. Before we get too excited about hypersonic speakers, let's note two major limitations of the current technology. First, though quality is improving all the time, they still have poor bass response: below 400 Hz their response is poor. This makes them unable to deliver music at anything much outside AM radio bandwidth. Second, their quality is limited: nobody could ever describe them as hi-fi speakers, because there is implicit distortion in the modulation/ demodulation process they use. This is particularly noticeable when using pure tones with few harmonics, such as sine waves.

In our experience, hypersonic speakers are most effective where the signal is relatively high frequency and quality is not the prime concern. Human voices are acceptable, with women's voices greatly preferable to men's. Classical music if well chosen to have little bass and lots of harmonics, for example string quartets or midrange piano or guitar. Birdsong and water are ideal content. For anything else, arrays are more effective, though there is a trade off in their directionality, which is less than that of hypersonics.

With any highly directional sound it's also important to be aware of reflections. If uncontrolled, these mean that sound bounces all over the place. If well managed, they can be highly effective because the sound appears to come from any surface it hits. In the branches of Helm Bank in Colombia we mounted hypersonics above the 'brand beacons'– graphical abstractions of a tree with metal leaves; the sound of a tropical rainforest seems to be emanating from the leaves themselves. Sound beams can also bounce off mirrors, refracting and reflecting to create sound pools exactly where we want them. Clearly it's important to plan both content and environment to maximise the effect of directional (and especially hypersonic) loudspeakers.

At the other end of the spectrum from focused sound, a highly diffuse sound source is needed if we want to create a room soundscape without any pronounced hotspots. This is where surface transducers can help.

A transducer is simply a device that converts energy (or a signal) from one form to another. All loudspeakers are transducers, as they convert electrical energy to sound energy, but traditional speakers use specially shaped cones to do this efficiently.

Surface transducers do this in a slightly different way: they convert electrical energy to vibration of a rigid piece of material. In air their output is hardly audible, but when such a transducer is attached to a surface like a piece of wood or metal, the whole resonator becomes the equivalent of the diaphragm in a traditional loudspeaker and sound is heard from all over it. This approach has been tried with varying success over the years, but it now has three interesting applications.

The first and oldest application is invisible hi-fi: high quality flat-panel domestic loudspeakers that can even form part of the walls of a room, removing the clutter of wires and speaker cabinets. Licensed by developers such as Sound Advance Systems in the US and NXT in the UK, these systems use custom-made flat diaphragms made of special materials to deliver excellent quality.

The second, relatively recent application is turning common flat materials into loudspeakers by attaching transducers (or more precisely exciters) to them. The best resonators are stiff sheets of resonant material: glass is excellent, but wood, metal and even plasterboard work. This means that tables, floors, walls and windows can all become effective loudspeakers, with obvious benefits for home cinemas, office presentation rooms, ATMs and advertising signage, all of which would gain from not having visible loudspeakers and cables.

Because these kind of surface transducers need to be in contact with a resonator, not the air, they are excellent for outdoor use: they can be bolted onto metal structures such as bus shelters, then covered with a vandal- and weather-proof metal case and left to get on with, for example, delivering classical music as an antisocial behaviour countermeasure. A conventional loudspeaker might be vandalised within days but such a surface transducer is almost indestructible. With economies of scale, they may be a very interesting alternative to centralised public address systems in large spaces such as major railway stations; fitted to hundreds of structures, and broadcasting locally, they would not echo like conventional speakers and the signal to noise ratio would be radically improved.

On a smaller scale, there are exciting possibilities currently being researched for laptops, mobile phones and the like: instead of using a tiny conventional speaker with a diaphragm of a centimetre or two, FSTs can use the whole screen as a resonator.

Hull-based FeONIC has been exploring the potential of this technology since the 1990s. First it launched a toy called the SoundBug which was originally sold through the Science Museum in London (and can still be found on the Internet if you want a good Christmas present for an inquisitive youngster), and more recently it made headlines with a product called Whispering Window, where entire store windows are turned into loudspeakers by attaching one of FeONIC's devices. When London department store Peter Jones trialled the product in 2003, it found that the number of people stopping to investigate the window in question went up by over 50 per cent.[41]

A word of warning to any retailer reading that number and getting excited: please note that the sound created emanates from the window in all directions, travelling inside the store as well as outside to passers-by. Also, this kind of intrusive sound must be handled with great care: it could cause serious annoyance and become counterproductive. However, where the sound is well designed according to the Golden Rules, there is lots of potential here.

Another interesting development that has great potential for large-scape installations is 3D sound. There is now a proven system that integrates the vertical dimension to deliver (for example) the sound of a mosquito flying around your head with uncanny realism. The brainchild of New York sonic artist, sound designer and innovator Charlie Morrow, the MorrowSound™ Cube features an array of nine loudspeakers: one at each corner of a cube, plus a subwoofer. This array can be installed in any size of room: for very high spaces, Morrow simply inserts one or more middle tiers of four extra speakers, like the jam in a sandwich. 3D sound like this has created stunning, high-profile installations in public and corporate spaces around the world, for example a 72-speaker soundscape for the Nokia World conference in Barcelona in 2008 and a 3D Brazilian rainforest soundscape installed in 2010 in SC Johnson Inc's Foster-designed Fortaleza Hall in Racine, Wisconsin, USA.

2.5 The SoundFlow™ model

At The Sound Agency we have developed a process that helps us to map, or to predict, the effects of sound on people. We call it SoundFlow™. It starts with five sets of drivers of sound, allowing us to extract all the aspects of any sound that will affect people. It considers four sets of filters that modulate the drivers and change the final outcomes, which again comprise four sets. Here's how the complete model looks:

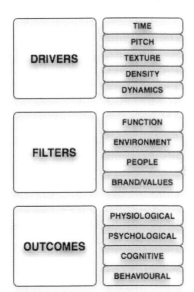

DRIVERS	TIME
	PITCH
	TEXTURE
	DENSITY
	DYNAMICS

FILTERS	FUNCTION
	ENVIRONMENT
	PEOPLE
	BRAND/VALUES

OUTCOMES	PHYSIOLOGICAL
	PSYCHOLOGICAL
	COGNITIVE
	BEHAVIOURAL

SoundFlow™ can be used either to audit an existing soundscape or to plan an optimally appropriate one for a space. In auditing mode, we start from the drivers, analysing the drivers that are currently applying, work through the filters in the same way, then estimate the current outcomes. This is often surprising, and it is in the rigour of the process that so much is revealed. Remember, people are generally unconscious of the sound around them and of its effects on them. In planning mode, we start at the bottom with the desired outcomes, work through the filters and arrive at rules for the soundscape that inform and even brief the creative process of actually making it, and the technical process of installing it.

Let's go through from top to bottom. There is so much to learn in this short journey.

Sound drivers

Time

This group includes all the time-related variables relating to sound, such as **tempo, rhythm, metre** and **duration**.

Tempo is the most powerful way of entraining physiological rhythms: faster tempos (at loud enough volumes) excite people, while slower tempos tend to induce relaxation, with powerful effects for business, as we'll see shortly.

Rhythm can also be powerfully effective. Specific rhythms, for example heartbeats or musical rhythms, may have associations that produce consistent effects. For example, I can't hear the drum intro to Stevie Wonder's *Superstition* without a touch of excitement and a slight increase in my heart rate. Other rhythms may use particular frequencies that resonate to create specific reactions in humans, for example tribal or shamanic drums (more on these below).

Metre is a building block of rhythm, a short, repeating group of emphases that create a characteristic combination. It's best known from poetry, but it exists in music also, and it affects us. Rigg (1937)[42] found that joy arose both from fast tempo and also from iambic (de-DUM, de-DUM) or anapaestic (de-de-DUM, de-de-DUM) metre. Sad feelings were evoked by trochaic metre (DUM-de, DUM-de) and by slow tempo.

Sounds with regular rhythms are predictable, making them easier to understand and thus to suppress (for example we can quickly learn to ignore a printer churning out 500 sheets of paper) – though there is mental work involved and doing this all day would be very tiring. Irregular, arrhythmic sounds can drive us mad – just think of a dripping tap – unless they become frequent enough to become stochastic and homogeneous, like rain.

Rhythms are not just found in music: they exist all around us. Most buildings have their own rhythms; so do most people. As always, the trick is to be conscious of them and their effects. It's all too easy for conflict to arise when a person with a snappy, brisk tempo tries to work with another whose tempo is laid-back. Sometimes the vibrations are just to far apart for entrainment to bridge the gap. This can be a serious issue in team working. I have seen drum workshops work wonders in literally bringing a team into sync; chanting can achieve the same outcome. Sound can be a direct tool in people management as well as creating better environments.

Pitch

This group includes all the frequency-related variables, such as pitch, harmonics, mode, melody and harmony for musical sound, or the predominant frequency of a non-musical noise. These elements can have strong effects, mainly on our emotions.

Certain principles are well known: for example, the major mode is almost universally perceived as cheerful, while the minor is more melancholy. A mode is essentially a defined palette of seven notes used to make music from. There are many modes, and they all have a different feel. Plato and Aristotle spent much time debating the effects of music composed in each of the ten Ancient Greek modes on its listeners, and the twelve Church modes of Mediaeval and Renaissance times (different from, but confusingly sharing, many of the same names as their Greek ancestors) were deliberately used to create different emotional responses to plainsong and Gregorian chant.[*]

If modes can create moods, so can melodies and harmonies. Most Western music is a journey from the tonic to the dominant and then back again, and the completeness of a song or a movement of a symphony that

---

[*]   Harmonics expert and overtone singer David Hykes has now rationalised the partial mode list of olden days, creating a full set of all modes. He says: "I use in Harmonic Chant and the Harmonic Presence work what I call the 32 Harmonic Mode Mandala, a full set of all possible seven-note modes, with the tonic and dominant unchanging. This all-inclusive system brings together a far greater set of modes than in Europe's past, and basically serves to show the universality of musical scales in the same way that my work with overtones shows the universality of harmonics." For more on Hykes, see www.harmonicworld.com or buy the CD Hearing Solar Wind.

is so structured is satisfying. By contrast, listening to Ornette Coleman is not so comforting: better descriptions would be challenging, stimulating or even intimidating.

The research in this area is long established. Hevner (1936)[43] proved the major/minor mode effect was real, and also found that simple harmony evoked happy and serene feelings, while complex harmonies were found to be exciting, vigorous or sad. In the study already quoted above, Rigg (1937)[44] found that joy arose from ascending fourths in the melody, major mode, simple harmony and staccato notes. Sad feelings were evoked by descending minor seconds in the melody, minor mode, legato phrasing, dissonance and low register.

Pitch may also affect our energy levels, according to Dr Alfred Tomatis, French sound guru and founder of the eponymous Tomatis Method that is used worldwide to help people with hearing disorders, learning disabilities, depression and other problems. As well as concluding from his work with opera singers that the voice cannot produce what the ear cannot hear, he noticed that we have four times more nerve cells in our cochlea that respond to sounds over 3 kHz than those responding to frequencies below this.[45] He believed that as a result higher frequencies (including harmonics) charge up our neural system, making us more alert, while large amounts of low frequencies overload us and cause the opposite effect.

It's interesting to consider the different effects on people of Mozart (lots of high frequencies) and rap (masses of low frequencies). Building on Tomatis's original propositions and a study by Rauscher, Shaw and Ky in 1993 at the University of California at Irvine[46], US author Don Campbell has created a major industry around the 'Mozart Effect', centred on the finding that listening to Mozart can boost people's performance in spatial-temporal reasoning tests (albeit only temporarily: the original paper found that the effect wore off after 15 minutes). The jury is very much out on this thesis. Nobody has yet been able to replicate the results of the original study, and it has been widely misinterpreted, much to its authors' dismay, as saying that Mozart makes people more intelligent. However, the success of the Mozart Effect as a business despite doubters and lack of hard evidence may be due to an intuitive resonance we all feel with something about the proposition that higher frequencies stimulate cognition.*

Compare and contrast this intuitive sense of elevation with the effects of lots of bass sound, as for example in rap or reggae music. Given where rap and reggae come from – the slums of New York and Kingston, Jamaica respectively – it rings true that their massive low frequencies offer an alternative way of zoning out, escaping an unpleasant reality. It is unlikely that such music would have arisen in the Vienna of the eighteenth century even if they had had the technology, and it seems equally unlikely that much intricate, high frequency work will come out of the deprived areas of Detroit or Los Angeles. I remember seeing people actually sleeping inside huge bass bins at club gigs in my playing days. I have no doubt they were sleeping something off, but you could never imagine someone doing that inside a high frequency horn.

I have a feeling that Tomatis may be right about sound being a kind of food for the nervous system, and that higher frequencies do charge us up mentally in some way. This may be another factor in the popularity of birdsong, which has negligible amounts of low frequencies. Perhaps the birds' dawn chorus is Nature's way of refuelling us for the day ahead.

Texture

This group includes all waveform-related variables, such as **timbre**, **texture**, **frequency profile** (known in physics as spectral density over time as depicted in sonograms or spectrograms), **instrumentation** or **vocal characteristics**, as well as **style** in the case of music.

Timbre is the property that allows us to perceive differences between notes of the same pitch and loudness. It has a lot to do with the structure and intensity of the partials (harmonics) of the tones concerned.

Texture is a more general term we use for complex soundscapes made up of many different individual sounds, with a meaning very similar to the one it has when applied to surfaces: when we discuss texture, we are describing how a soundscape *feels*. Some of the describing words will be: rough, smooth, hard, soft, sharp, harsh, flat, soothing, diffuse, and so on. In music, texture is determined mainly by instrumentation, as well as

---

* Campbell has gone on to stimulate a lot of work in the broad area of sound and health, which I believe is rich in potential, so his empire is no longer based on the original single proposition that Mozart makes you smarter.

the style of playing. Just listen back to back to the Frank Sinatra and Sid Vicious versions of *My Way* to experience the power of texture in music – same song, completely different effect. The overall style of a piece of music must also be considered, since most people have well-formed preconceptions of what they like and don't like. However good it is, bebop jazz is unlikely to work as a soundscape in a young girls' accessory shop simply because the style is wrong.

The frequency profile of the sound across time is very important when dealing with background sound such as machinery or traffic. If the sound energy peaks at levels where the human ear (or specifically that of the listener) is most sensitive, the sound may create a negative effect, causing stress, frustration or upset, because it interferes with our concentration or our efforts to listen to foreground sound.

Texture is a key determinant of appropriateness and acceptability, because it is the prime determinant of pleasantness.

Density

This group includes all density-related variables, such as the number of **themes** or **events** and the degree of **variability** or **contrast** that we discussed as one of the three Cs. These factors are prime determinants of the effects of sound on cognition – in other words, how distracting it is.

In our discussion of listening, we explored both pattern recognition and differencing, the two techniques our reticular activating system (RAS) likes to use as filters on sound. The RAS will generally send auditory events to the cerebrum for analysis unless the pattern is well known and deemed unworthy of attention – for example the background noises in our house – or unless it is repeating and unchanging, in which case it stops being sent as soon as the RAS is confident there is no new information content.

This makes us very good at suppressing sounds that are constant, like hums: once we have recognised their pattern, we cease to be conscious of them unless they change. We also do this with sounds that have repeating patterns. The hiss of an air handling system, the whirr of the cooling fan on a computer, even the steady repetitive sound of trance music – all these we can ignore after a time.

However, sound that the RAS judges to have new information content is a different matter. Complex melodies or harmonies, unpredictability, large numbers of audio events and, most of all, the human voice – all these are guaranteed to be stamped 'for urgent attention' by the RAS.

This holds true up to the point where any of these sounds become stochastic. A dripping tap is unbearable, but steady rainfall is soothing. One person talking is impossible to ignore; a whole roomful of people talking becomes possible to work over.

Dynamics

This group includes amplitude-related variables such as **loudness, dynamics** and **waveform**. In general, louder sound simply has larger effects. Dynamics (the pattern of variation in amplitude) are important too: because our primal self-preservation reflex works entirely at sub-cortical level, sudden loud sounds with short attack times are guaranteed to grab our attention, and in many cases they trigger a series of physiological reactions designed to prepare us to fight or flee. Waveform means the way a sound's energy is distributed over time – the ADSR profile we reviewed in Part 1. This too can have an effect: for example, sounds with short attacks, such as bangs or crashes, are more startling than sounds that build slowly.

Some musicologists (and notably the American philosopher Susanne K Langer[47]) suggest that the dynamics of music have a structural resemblance to the forms of human feeling. Langer writes: "Every artistic form reflects the dynamism that is constantly building up the life of feeling."

The prominent psychiatrist and specialist in child development Daniel Stern takes this a stage further, proposing that human beings are emotionally sensitive to 'vitality effects' – a set of elusive qualities related to intensity, shape, contour and movement and best described in terms that are more commonly applied directly to music, such as crescendo or diminuendo. Stern says that infants "take sensations, perceptions, actions, cognitions, internal states of motivation, and states of consciousness, and experience them directly in terms of intensities, shapes, temporal patterns, vitality effects, categorical effects, and hedonic tones."[48]

Stern claims that vitality effects are conveyed directly by music and dance, and these responses last through life – so according to this theory we are all innately **musical from birth and it is dynamics that create** the most potent effects of music on our feelings. The conclusion that we are born musical is one we have already come across from authors such as Mithen, Ball and Levitin in Part 1 where we discussed music as one of the key types of sound. It seems that many scientific roads arrive at the same place. If it's true, our innate musicality does help to explain why sonic dynamics affect us so powerfully.

Filters

Function

What is this space for? What are people trying to achieve in it? What are the intentions? How can sound help? What effect is sound currently having?

It seems obvious that a library needs a different soundscape from a bookshop, which in turn is very different from a record shop – but it is amazing how often this aspect is not considered, especially in relation to time.

Function very often changes through the day, or across days of the week. Bars are prime culprits in failing to consider the way their customers' intentions change throughout the day, with the result that many of them purvey a homogeneous soundscape of loud music from 10am to 2am, regardless of whether people are trying dance or to have quiet conversations. It doesn't have to be this way: we designed a musical soundscape for the bar in the London InterContinental Hotel Park Lane that moved from 'Sunrise' (gentle, ambient music from late morning to mid-afternoon) to 'Sunset' (laid-back lounge and deep house grooves for the cocktail hour) to 'Sundown' (classy music with energy and a buzz to it) for the evening session.

Environment

What are the acoustic properties of this space? Can we improve them? What sound will work in this acoustic context, and what won't? What other sound is going on in the space? How good is the sound delivery system?

It would be crazy to play dance music or to hold an auction in a cathedral. The acoustics would blur and render meaningless the sound in either case. Conversely, Gregorian chant would sound flat and lose most of its charm in a dry recording studio. Any pronounced acoustics of a space will be crucial in determining what sound we can deploy there. In many cases, remedial work needs to be carried out to make the space fit for purpose, because the designers didn't take acoustics into account when they decided on wood, glass and metal everywhere. Cafés and restaurants are prominent sinners in this respect. Just visit your nearest Starbucks and listen to the profile of the reverberation. If a space is going to play host to all that sound, its acoustics should be designed to minimise the discomfort and interference with speech. Sadly I have yet to find a Starbucks where this is the case; in fact, the typical acoustics, far from absorbing the painful frequencies, emphasise them. Another example is the noisiest restaurant I know: Kensington Place in London's Notting Hill. The food is delicious, but at peak occupancy it is impossible to converse at less than a bellow, and even then it's hard to understand your companion. I have no doubt you can think of many such examples from your own personal experience. With a little help from a professional acoustician they would all soon sound completely different.

Equally important to consider are any other sounds or noise sources in the space (for example the sound of people shopping or eating or doing whatever they do in the space) or intruding sounds that you can't remove, such as road traffic noise.

Finally we must think about the system that will be delivering the sound. If we've determined that a funky clothes store should have loud music playing for its twenty-something customers, we must ensure that the system is up to the task. Few sounds are as unpleasant as distorting amplifiers or loudspeakers. Speaker placement is also critical, particularly in the case of systems with a large subwoofer and small satellite speakers, where the sub most be very carefully positioned if it is not to intimidate people unfortunate enough to be close it with massive bass.

There are so many good quality and cost effective amps and speakers on the market that there is no excuse for some of the systems we find in shops, with tinny ceiling-mounted speakers that simply say "we are cheap" to the customers. As already stated in the section on sound technology, do not skimp on your sound delivery system: you are better off with no sound at all than with something that produces low quality sound.

People

Who are the people in this space? What is the demographic and psychographic profile of this group? What are their tastes, cultures, tribes? What sound will they like and what will they dislike?

Again it seems obvious to create sound that your customers are going to like but in my experience, making assumptions about this is dangerous: investigation does not always yield predictable results.

First, it's not always obvious to the people creating the sound who the customers actually are. When we did an audit for a well-known children's toy store chain, we found that most of their stores were playing the tapes they sold: some nursery rhymes, but mainly kiddie-pop music, because the staff found it cheered them up and gave them energy. On the surface this seems entirely appropriate – until you consider that the children are not in fact the customers. It's the harassed, frazzled mums with two children in tow at the end of a long and tiring shopping trip who have the credit cards. The soundscape in these stores was chaotic, full of toy noises (children are encouraged to play with the stock) and banal, jolly kiddie-pop. This was not pleasant for the average customer, who would attempt to buy what they had to and then escape as fast as possible. We trialled various sound and found that slow, serene classical music soothed the atmosphere; dwell time increased as a result and so did sales – by eight per cent. The staff had mixed reactions, but I would suggest that good training and communication of the effects on their bonuses would solve this.

Second, it's vital to remember that what people say is not always related to how they behave. Peer pressure, preconceptions and simple bigotry will often come into play when you ask people about sound, and in particular about music. Your customers may say they prefer to listen to fast-paced dance music, but if the effect is to speed up their shopping and cause them to spend less time in your store, is that the most appropriate soundscape to choose? As we've said before, it's what people do that matters, not what they say.

Having said that, preferences are an important part of this filter. Herrington and Capella (1996)[49] found that the time people spent in supermarkets was related to the degree to which they liked the music – and that longer shopping times were related to higher expenditure.

Sullivan (2002)[50] found that popular music in a restaurant led to meal times twice as long as when unpopular music was played, with predictable effects on spending.

A negative relationship between sound and people has been used successfully too. Classical music has been played in many public places as a deterrent to groups of young people whose behaviour or presence was perceived as threatening or otherwise upsetting to the bulk of the customers. Newcastle's Tyne and Wear Metro heard of the idea from Canada and installed speakers playing classical music to deter groups of young people who were swearing, smoking and abusing passengers. "It has completely eliminated the problem," said company spokesman Tom Yeoman. "The young people seem to loathe it. It's pretty uncool to be seen hanging around somewhere when Mozart is playing." Similar schemes have been trialled with success in the UK by other transport networks, including London Underground, which rolled out classical music to over 100 stations after successful tests.

Clearly such schemes are not in themselves a solution to antisocial behaviour: they are simply moving it somewhere else. I don't think there is any danger of us arriving at a world universally veneered with Mozart string quartets to damp down behaviour. That would be unacceptably Orwellian, reminiscent of the mind control exerted by the controllers of the Village in the classic TV series *The Prisoner*. I do not ever advocate using sound to manipulate people, but I do believe we can usefully move towards a world where soundscapes are thoughtfully designed to support people in doing what they want to do. At the same time, it's useful to know that there is a relatively harmless extra tool to defuse local hotspots of antisocial behaviour, improving the quality of life of the many at the expense of the indulgence of the abusive few.

An excellent example of highly targeted sound which really understood its audience was the Ogilvy-designed Fanta Stealth Sound System. This mobile app used frequencies above the hearing of older people but perfectly audible to teenagers to let its users communicate with each other while in public places like classrooms. Human League founder and sound designer Martyn Ware created a set of high-frequency sound tags for simple messages like 'yes', 'no', 'cool' and 'run for it', which were transmitted at the touch of a button, while any adults in the room stayed completely oblivious to the banter. The application had more than half a million downloads across Europe.

The important thing, as always, is to be conscious of the effects of sound, and specifically in this case of the relationship that will form between the people in a space and the sound they experience there. After all, they are the only reason the sound is there. You must ensure that you understand their demographic profile because age, gender, income and location can all have an effect on the way they relate to sound. Even more important is to understand their psychographic profile: cultural conditioning, values, fashion, attitudes and lifestyles can all change the way a sound is received.

Brand/values

What is the brand behind this space and what are its values? (If there is none, what are values associated with this space?) What expectations do people using this space enter it with? What sound fits well into this emotional/psychological context, and what doesn't?

Context is massively important when considering sound. Women seeking, say, socks or gloves enter a Zara store with different expectations from those with which they enter a Tesco or Walmart. What's acceptable or appropriate in one (in terms of range, service, facilities, price, environment) may not be so in the other. This is the power of the brand, and that context is vital to consider when thinking of sound. We'll be looking at brands and sound in more detail in the next section.

Of course it's not just brands that have associated values. Public buildings, religious spaces, transport termini – they all have histories of personal and community experience that must inform any soundscape that's designed for them.

Outcomes

Now we're going to look in detail at the four ways in which sound affects people: physiologically, psychologically, cognitively and behaviourally.

Just before we do, I want to emphasise that most of this is unconscious. We are so used to suppressing and ignoring the sound around us that we have lost the skill of recognising these effects as they happen. We are a species in denial of the power of sound to affect us. That's why I am generally so reluctant to suggest market research to clients (as in "what do you think of this sound?"): sound is always having an effect but if you ask people what they think, they will have no idea what that effect is. I have experienced some truly dreadful soundscapes in shops and it is rare to find that they have had even a single complaint. *That doesn't mean there are no adverse effects on customers and sales; it just means people are not conscious of them.* If this book achieves one thing only, I hope it is to move us out of this denial and into recognition of the enormous, varied and wide-ranging effects that sound is having on us all.

Physiological

Largely through entrainment, sound can affect all the major rhythms of the human body, primary among them our heartbeat, breathing cycle, hormonal secretions and brainwaves.

The average resting healthy human **heartbeat** has a frequency of between 60 and 80 beats per minute (bpm). In high fever, during strenuous exercise (including sex) or in states of high arousal, excitement or fear, this rate can rise to over 200 bpm. It comes as no surprise that we find music with tempos below 80 bpm calming, while dance music at 150 bpm stimulates and excites us. There is a feedback loop in each case: we associate the tempo with a state and start to move into it, while at the same time the louder the music is the more it entrains us physically and the more our heart rate changes to match its tempo.

The heart beats in three time (lub-dub-pause, lub-dub-pause), a rhythm shared by the resting breath (in-out-pause) typified by a sleeping person, and also quite possibly the way you are breathing right now. Three-time has soothing connotations the world over as a result. Even the fastest three-time of the Viennese waltz is uplifting and blissful rather than violent: spinning, not stomping.

Two-time or four-time are associated with our other breathing cycle, the panting of exertion or stress, and also with our two feet marching in time. (It is interesting to speculate what kind of dance music we'd be moving to if we had three legs.) That's why all military music and dance music are in two-time or four-time.

Going back to **breathing**, our resting breathing cycle has a frequency of around 6-7 cycles per minute (cpm). It was American sound expert and author Joshua Leeds who pointed out to me several years ago that this is why we find gentle surf so relaxing: its cycle often mirrors that of our sleeping breath. If you have trouble sleeping, try putting on a recording of gentle surf, ideally at 6-7 cpm in three-time – or any sound with those characteristics – and I predict your problem will be greatly eased.

We have many more subtle cycles running in our bodies, many of them involving **hormones**, and there is a whole class of science called chronobiology now studying them. Women have their monthly menstrual cycle – possibly the only human 'infradian' cycle (lasting longer than a day), unless the theory of biorhythms is ever substantiated. We all have a basic 'circadian' rhythm of roughly 24 hours, kept in time by daylight, that drives many hormonal sub-rhythms like a cam belt driving all the separate cyclical events in an engine. Hormones are secreted at appropriate times to wake us, deal with food, excrete and so on. Jet lag is what happens when all these circadian rhythms are out of sync with our new time zone. Finally we have multiple 'ultradian' rhythms (shorter than one day), such as our 90-minute rapid eye movement (REM) cycle during sleep, or the three-hour cycle of growth hormone production.

Sound can interrupt or alter the running of this amazing machinery – in particular by operating in its primal function, which is to warn us of danger. We are not so modern that a sudden sound does not cause an immediate release of adrenaline and cortisol, our fight/flight hormones. By the time we realise it was just a bus or a car backfiring, it's too late. As we go about in urban environments, we are being dosed with these hormones over and over again because of the noises we encounter. The effect of these hormones is to increase our blood pressure and our blood sugar levels, preparing us for action; cortisol also inhibits our immune system and reduces bone formation. These are stress hormones, and their continuing dosage is known to result in muscle wastage, hyperglycaemia and brain damage, particularly relating to impaired learning; also the repeated triggering of our flight/flight reflex without any action following is believed to result in symptoms such as diarrhoea, constipation, and difficulty maintaining sexual arousal. Who knows how many of these effects are being felt by city dwellers without any causal link to sound ever being made? Much more research needs to be carried out in this area, and soon.

The other critical rhythm in our bodies is electromagnetic rather than physiological/mechanical. Our **brainwaves** are created by the electrochemical activity of millions of neurons firing in synchrony in ways that are at best only partially understood to date. In his book *Sync*, Steven Strogatz reports one interesting hypothesis, proposed by neuroscientist Christof Kock and his collaborator Francis Crick (the same Crick who discovered the double helix), who suggest that our whole level of consciousness is determined by the degree of synchrony going on in our brains.[51]

When we measure brainwaves we are measuring neurons firing in sync, often in very different parts of the brain. Usually there are several rhythms going at once. What each rhythm means continues to be the subject of great experimentation and speculation. There is a huge amount of scientific research into the relationship between brainwaves and mind states, and an equal amount of practical, experiential evidence coming from alternative medicine and also from the fringe. The most commonly held view seems to be that our mental state (particularly our level of alertness or mental activity) maps consistently onto the predominant frequency of our brainwaves – in other words, we can infer that a person exhibiting primarily brainwave x is in mental state y.

Brainwaves have traditionally been grouped in four frequency bands, all of them relatively low compared to sound waves. In this traditional view the highest frequency brainwaves are Beta (roughly 12 to 38 Hz, though there are no internationally agreed standards on these frequency bands – the exact transition points vary from one source to another). Beta waves are predominant when we are active and alert. Next rung down on the ladder is Alpha (roughly 8 to 12 Hz), found in resting, meditative states. Deeper into relaxation and we find the transitional state Theta (4 to 8 Hz), and below it the deep sleep state of Delta (0.5 to 4 Hz). There is in theory a further rung called Epsilon, but this is restricted entry; you have to be in suspended animation to get in here.

More recent research has discovered whole new levels of brain activity at much higher frequencies than Beta, sometimes manifested as interference riding on the slower, longer waves down below. Gamma (20 to 40 Hz) is thought to be associated with synthesising information from all over the brain to grasp and integrate whole concepts – what neuroscientists call the binding problem.[52] As for the even higher frequency activity at Hyper Gamma (up to 100 Hz) and Lambda (100 Hz to 200 Hz) levels, their role is as yet unproven. On the web there is plenty of speculation that they are involved in higher states of consciousness but it will be some time before the steady process of science has produced reliable answers.

It is important to record that there is still great debate in the scientific community about any simple mapping of brainwaves onto mind states. There are plenty of studies exploring specific states that conclude that increased levels of consciousness are not necessarily correlated with increased brainwave frequencies.[53] This is a complex field and any generalisations are to be treated with care.

Having said that, it's useful to have a clear summary of the current thinking from all sources, so I reproduce overleaf a table of all the known brainwaves.[*]

This is important information because researchers have found that brainwaves can be entrained like any other rhythm through what is called a 'frequency following response'.[54] The ability to persuade the brain to move from one state to another has obvious applications in fields such as education, where fatigue or boredom could be fought off by encouraging the brain to stay in Beta – or in medicine, where conditions such as insomnia or stress disorders could be ameliorated by moving the brain into Delta or Alpha respectively.[55]

Name	Frequency	Characteristics
Lambda	~200 Hz	Self-awareness, higher levels of insight and information. Tibetan monks who walk barely clothed for days through the snow have exhibited high levels of these. They are difficult to measure and little is known about them. They are carried on the very slow moving Epsilon Waves (<0.5Hz).
Hyper Gamma	~100 Hz	
Gamma	38 - 90 Hz	Important in harmonising and unifying thoughts processed in different parts of the brain. Combines different perceptions. Suppressed totally by anaesthetic. Found in all parts of the brain. Self-awareness and insight.
	40 Hz	The core frequency. Important in cognition, especially coordinating simultaneous processing in all parts of the brain. Deficiencies exhibit learning difficulties. Produced during hypnotic states.

---

[*]   I found this on the website of London hypnotherapist Ralph Price and reproduce it with his permission, but with the proviso that no sources are given for the claimed characteristics.

Beta	12 - 38 Hz	Wide-awake, alert, focused. Analyses and assimilates new information rapidly. Complex mental processing. Peak physical and mental performance. Cannot be sustained indefinitely otherwise exhaustion, anxiety, and tension result.
	15 Hz-18 Hz	Mid Range Beta. Neurofeedback training that produced alert behaviour, useful in depression cases.
	12 - 15 Hz	Low Beta. Also known as Sensory Motor Rhythm (SMR) – vigilance, reduced mobility, shallow breathing, less blinking, fixed attention and eye focus. Enhancing through neurofeedback reduces epileptic symptoms and has a calming effect (ADHD sufferers).
Beta-Alpha	12 Hz	Hyper-efficient in processing single tasks as it can focus on the details as well as the overall task at the same time
Alpha	7.5 - 12 Hz	Mental coordination and resourcefulness, relaxation. Alert but not mentally processing anything. Inward focus, calmness, at ease. Deep breathing and closed eyes can amplify alpha production. Peak around 10Hz.
Alpha-Theta	7.48 Hz	Primary ionospheric resonance (Schumann) frequency. Stimulates retrieval of memories from the subconscious.
Theta	4 - 7.5 Hz	Memory access, learning, deep meditation, sensations, emotions. The threshold of the subconscious, dreaming.
	6.2 - 6.7 Hz	Frontal Midline Theta – Cognitive activity, maths problems, sustained attention. Extrovert personality, low anxiety
	4.5 Hz	Shamanic trances, Tibetan mantras, Buddhist chants all use this frequency to access altered states
Theta-Delta	~3.5 Hz	Long term memory access
Delta	0.5 - 4 Hz	Deep sleep, human growth hormone release, low blood pressure, low respiration, low body temperature. No muscle movement – Reticular Activating System (RAS) shuts this down.
Epsilon	<0.5 Hz	The state Yogis go into when they achieve 'suspended animation' where no heartbeat, respiration or pulse are noticeable.

Actually, brainwave entrainment may not be so new: it has probably been a consistent human practice for thousands of years. The lower brainwave frequencies are well below musical pitch and fall into the frequency bands that can be played by drummers. This is true of the interesting Theta band, within which the frequency 4.5 Hz is widely associated with trance-like states. One interesting piece of work showed that drumming patterns with frequencies of 4.5 Hz do induce the brain to move into Theta waves.[56] This particular frequency of drumming (and/or chanting) has long been used in tribal and shamanic rituals designed to induce trance-like states. Perhaps we are simply rediscovering one of the oldest forms of entrainment.

The rather less intrusive modern mechanism proposed for brainwave entrainment is beat frequencies. If I play you a tone of 800 Hz in your left ear, and one of 1000 Hz in your right ear, you will hear the difference between them as a third tone of 200 Hz in the centre. No such tone is being played, but it will be as real as the other two to you. Now if I make the difference between the tones 30 Hz – sub-audible to many people, just discernable to others – we have a Beta beat frequency and, so the theory goes, your brainwaves get entrained to match it.*●

Beat frequencies were discovered by German experimenter H W Dove in 1839. If you're wondering how he did it without headphones, the answer is that he placed his subject in one room, rigged up a system of pipes to carry sound from the adjacent rooms on either side (otherwise completely sealed for sound) and then had tuning forks play simultaneously in the two adjacent rooms. One wonders how he ever got the idea to do all this! Today we can get the same effect somewhat more easily with a pair of headphones – ands this is the only way I know of beat frequencies working. I have encountered some claims of headphone-free beat frequencies, but I am highly sceptical.

Given the potential benefits of brainwave entrainment, it comes as no surprise that there are hundreds of CDs on the market offering everything from enhanced wellbeing to instant enlightenment through the use of beat frequencies – why spend years learning to meditate when you can get the same effect in 30 minutes, suggest the less respectable among them. But there is sufficient serious research[57] to indicate that this is an interesting field with much potential. The largest practitioners, such as the Monroe Institute, sponsor active research programmes and are substantial businesses. Most of them claim that beat frequencies can not only entrain the brain into desired states, but also create a state of whole-brain synchronisation, whereas normally the two sides of the brain are out of sync. This whole-brain sync is often associated with a state of peace, wellbeing and high levels of effectiveness, and certainly there are plenty of testimonials to this effect from satisfied customers. The thesis resonates with the scientific work reported above on sync and

---

*● You can hear beat frequencies on the website.

consciousness, making this another interesting aspect to watch.

For Japanese interactive education specialist Cerego, The Sound Agency created sound 'skins' with beat frequencies underneath other programmed material, such as music or natural sound. The beat frequencies are in Beta range, with periodic dips into Theta: it is claimed in the beat frequency community that 15 minutes of Theta resets the brain's potassium and especially sodium levels, which become depleted after lengthy Beta activity, causing mental fatigue. I have found no scientific verification of this but again it seems intuitively plausible: we all get mentally fatigued and we do reset after a rest or a sleep.*●

Returning to our whole body and its response to sound, if in doubt about the potency of sound to affect us physiologically, consider the experience of New York Times reporter Marshall Sella when he encountered ATC's militarised application of hypersound:

> Norris prods his assistant to locate the baby noise on a laptop, then aims the device at me. At first, the noise is dreadful -- just primally wrong -- but not unbearable. I repeatedly tell Norris to crank it up (trying to approximate battle-strength volume, without the nausea), until the noise isn't so much a noise as an assault on my nervous system. I nearly fall down and, for some reason, my eyes hurt. When I bravely ask how high they'd turned the dial, Norris laughs uproariously. "That was nothing!" he bellows. "That was about 1 percent of what an enemy would get. One percent!" Two hours later, I can still feel the ache in the back of my head.[58]

Without going to these extremes, the sound around us all the time is continuously affecting all the major and minor rhythms of our bodies either through entrainment or deep reflex responses – and most of these effects are non-conscious.[59]

Psychological

I am sure there's a piece of music I could play you that would immediately make you wistful, even melancholy – and another that would just as quickly get you smiling, excited and energised.

Music is the first sound most people think of when they think about sound and psychological state. It has an uncanny ability to communicate

---

*● There is an example of some of the Cerego work on the website, including beat frequencies. See if you can hear them under the water and gongs.

emotion, and despite many studies nobody really knows how. The combination of pitch, movement, harmony, melody, rhythm, tempo, density and timbres is a rich and complex brew with so many drivers involved, and so many interrelationships, that we may never understand exactly how music works. We just know that it does have strong psychological effects.

It's not only musical sound that affects us in this way: voices, natural sound and noises can all alter moods and induce emotional states; arousing, stressing or calming people.

Some examples will show just how powerful the psychological effects of sound can be. Let's start with war. Where they can, soldiers march in rhythm, which fosters confidence and a feeling of strength (we are all together). Contrast the virile tramp of the victors with the arrhythmic shuffle of columns of POWs in defeat. In the same way, marching drums, bagpipes and the massed shout of an army charging are sounds that are designed to boost our courage and sap that of the enemy.

Some of these combative sounds have found their way into modern sport. The sound of 80,000 voices roaring when their team scores is an unforgettable and almost overwhelming experience; football chants are part tribal bonding and part territorial defence, and they can change the whole feeling of a game, lifting the team on the pitch and pulling them through to victory when their heads drop. (They are also possibly the last form of real folk music in the UK, coming from the streets and being constantly invented and reinvented by unknown troubadours somewhere in the crowd.)

In love, too, sound plays a vital role. Love is the largest topic on which music dwells by far: music appears to be uniquely effective in communicating both the extremes of love and loss. We slow dance, we have 'our song', we make compilations of romantic music for our beloved, and (if we are blessed with the capacity) we even serenade them; in all these ways sound is reinforcing as well as transmitting our feelings. During the act of love itself, for many people, wordless sound (metalanguage) is the primary way they communicate their feelings to their partner.

Human voices express a vast range of emotions, and this sound transmits those emotions to those listening, changing the way they feel. It's hard to hear someone in pain, or in distress, without feeling with them. The reason the sound of crying babies was so high on that list of most hated sounds in the last section is that we are psychologically (and probably genetically) programmed to look after children, so their discomfort affects us deeply and instantly.

Sound creates some of these potent psychological effects by a physiological process, when making or receiving sound changes our physical state as we've described above. Other effects are created by association – the meaning we give to a sound. It's fundamental to our process of perception that we associate with sounds whole packages of values, preconceptions, experiences, expectations and memories. In Pavlovian style, we react to the whole package of associations, not just to the sonic stimulus itself.

Sound is an excellent anchor for memories, and it can often be the case that hearing a long-forgotten sound will bring back a scene more vividly than a photograph ever could. When veterans of D-Day went to a special screening of *Saving Private Ryan*, many of them were overcome with emotion during the harrowing first half hour, which depicts the Normandy landings in graphic detail. It wasn't the sights that brought it all back to them: it was the sounds, which Steven Spielberg had researched carefully and which were faithfully reproduced for the first time ever in a war film.

Earlier we discussed how humans use pattern recognition to make sense of the sound around them. We do this precisely by associating meanings with patterns: we react to our own ring tone and not other peoples'; we start at the sound of our own name being shouted; we recognise and respond to the sounds that are designed to warn us (traffic signals, car horns), call us to action (alarm clocks, phone tones, door bells), change our mood (music).

Many natural sounds have strong associations. Earlier I wrote about my conviction that we're pining for wind, water and birds (WWB); I have installed enough birdsong to know from experience that people find it restful in general, because of the association at deep species level of birds singing with safety. Running water is the sound of life, and I believe we love to hear it largely because of the associations with survival, growth and fertility. The same applies, though less powerfully, to the sound of moving air.

Murray Schafer coined an excellent word – schizophonia – for the modern experience of sounds being experienced out of context, for example a recorded concert or a voice on a PA system. I think it also describes the experience of seeing something without hearing the appropriate sound – for example sitting in a hermetically sealed office and watching the rain. I have often imagined installing a system that plays the sound of rain inside such an office when rain is detected outside. I have a feeling it would be healthy for the people inside to be reconnected with the sound of the world.

As always, the human voice creates some of the most powerful associative effects of all: what we read into one tiny nuance in our spouse's voice can completely alter our emotional state – for better or worse!

Advertisers understand this process of association well, and much of the effort in music selection for TV ads is centred on trying to find or create music and voices with exactly the right positive associations.

As well as affecting us psychologically by association, sound also gets to our feelings and moods more directly, through our physiology. Many of the physiological processes described in the last section have automatic psychological effects. A fast heartbeat means that we are in some way excited – and it certainly precludes us from feeling calm, laid-back or meditative. Hormone secretions change our whole state, not just our physiology: those cortisol shots make us more alert, and bring with them a short charge of fear to jolt us out of complacency. When we make sound ourselves, there is an effect: it's impossible to scream or shout and remain emotionally calm.

Music can bring all these effects together to shift our psychological state. Several studies have shown that musical texture can change people's attitudes: in a soundscape of music perceived as pleasant, people are more willing to help[60], to volunteer[61], and to entertain sales approaches[62] than in a soundscape of music perceived to be unpleasant. In some places, textures themselves can be provocative because of their associations: fife music can stir anger in the Celtic areas of Glasgow and the separatist communities of Northern Ireland because it is associated with loyalist marching bands and the Orange Order. When English footballer Paul Gascoigne celebrated a goal for Rangers by miming the playing of a flute the Celtic fans were violently enraged because of this association.

The sublime effects of uplifting music are something we have all experienced. Of more concern lately have been the negative effects of heavy metal and rap music, which combine tempo, rhythm, pitch, texture, density and sheer volume to create a profound effect on their listeners and are strongly suspected of creating unhealthy psychological and attitudinal states, including misogyny, aggressiveness and depression, and of causing a range of antisocial and damaging behaviours such as violence, crime, self-harm, cult membership and even suicide.[63] The growing evidence underlines how careful we must be when thinking about deploying music in a public space. The decision is not trivial and its effects can be profound.

Music with lyrics predictably has even more potent psychological effects, because the words usually crystallise and make explicit the emotional content of a song and thus remove much of the scope for (mis)interpretation. At one end of the spectrum, wistful or sentimental lyrics work with all the musical elements we've already discussed to shift our mood towards romance or melancholy; at the other, violent and aggressive lyrics in rap or heavy metal music are clearly instrumental in creating the effects noted above. Lyrics can cause offence so easily that it's usually safer in commercial situations to deploy instrumental music. I have been embarrassed in the past by the offence caused by an obscene expletive in the middle of a long track in the middle of a long playlist of lounge music. It's important to ensure that all tracks are safe, by which I mean that they do not include obscenities, political, religious, sexual or other material that is likely to offend people.

Psychoacoustic theory on exactly how music affects us in these ways is somewhat limited. It concentrates on the concept of arousal. As we've seen, music stimulates activity in the brain's reticular activating system (RAS); Berlyne[64] claimed that the fibres of the RAS pass through both the pleasure and pain centres of the brain, and that moderately arousing music causes activity in the former but not in the latter. This theory – that mildly arousing music make us feel good – is behind much of the playing of jolly, up-tempo music in stores. I recommend careful testing because this contradicts the stronger result that slow music creates longer dwell time. I suspect that many stores are speeding up people's progress so that even if they do feel mildly aroused (and even if this does make them feel good – which is unproven) they spend less time and money there.

One other psychological effect that's been researched is the relationship between music and time. There is conflicting research on music and our perception of time: some papers claim that people think time is passing faster when they are exposed to an unpleasant (disliked) sound and other papers claim the exact opposite. We need not consider this for business applications until there is a clear consensus. However what is fairly clear is that people are willing to wait longer when listening to music, either on the phone or in person.[65] Music on hold really does work. But please read the section on music on hold before you rush out and set yours up!

Cognitive

Sound can aid or impede concentration, focus, communication and mental processes, particularly manipulation of symbols in short-term memory. This clearly has major implications for people's productivity. There is now a substantial amount of research that maps in detail the ways in which noise adversely affects people at work, particularly for the office environment. We now know that noise has negative effects on people at work; that it is related to reduced job satisfaction and dislike of the environment; and that it increases stress among employees.

As we saw in the section on noise, its effects are made worse by the three C's: lack of control, changes in the level or texture, and the amount of conversation it includes.

The starkest calculation of the results shows a staggering cost to business of office noise. Banbury and Berry (1998)[66] found that office noise, and particularly that including speech, reduced cognitive performance by two thirds. We will review this research in the section on staff spaces.

Behavioural

The simplest effect of sound on behaviour is that we tend to move towards pleasant sound and away from unpleasant sound if we can. This simple mechanism, driven mainly by the texture and loudness of a soundscape, is vital to consider, and it's the one that is causing many shops to lose business every day. The prime offenders are supermarkets, whose soundscapes are usually somewhere from mildly to extremely unpleasant. The effect is to encourage people to leave as soon as possible.

This is not a tempo entrainment issue: it's a texture effect. Based on our experience in the field, I believe that many supermarkets could increase their sales by at least 10 per cent simply by dealing with the noise they are unconsciously, uncaringly inflicting on their long-suffering (but equally unconscious) customers.

All the other sound drivers affect behaviour, and the effects can be important for business. The most famous study of tempo and entrainment in retail is that of Milliman (1982)[67] who tested three soundscapes in a medium-sized US supermarket: no music, slow music (<73 bpm) and fast music (>93 bpm). He found that the pace at which people moved through the store was affected by the tempo of the music – and that as a result people spent 38 per cent more in the slow-music soundscape than in the fast-music soundscape. The likely explanation for this is intuitively reasonable: moving more slowly, people are more likely to notice things and therefore to make discretionary purchases, as opposed to whipping through their list.

Most retailers I know are very interested in a 0.38 per cent increase in sales, let alone 38 per cent. Perhaps this research is just not widely enough known, but it is amazing that the soundscapes in so many shops feature only fast-paced music, which will encourage people to shop and run. Every retailer knows the correlation between dwell time (how long people spend in the store) and sales. Sound has a major role to play in increasing dwell time, in the simplest case simply by creating more pleasing and relaxing soundscapes for customers.

Milliman (1986)[68] repeated the exercise in a different setting, this time a restaurant. The results were consistent with the first study. Fast music caused people to eat faster and spend less on drinks as a result; the restaurant's margin was 14 per cent higher in the slow music condition. This has to be tempered by the time element because a very busy restaurant might prefer to move people through faster, even at the expense of that kind of drop in margin, if it allowed another entire seating to be served during a session. The value of time in both contexts will be different for a restaurant with a standard, relatively homogeneous bill of fare and not many extras (for example McDonald's) compared to one with a large menu and an extensive wine and *digestif* list. In the former case, fast music might be ideal to increase the number of covers, while in the latter profits probably rise in a straight line with the length of time people spend over their meal. Other studies[69] have backed up these findings for catering: it seems clear that fast music causes people to both eat and drink more quickly.

Behaviour is affected by the pitch drivers as well, through familiarity and predictability of melodic or harmonic structures. Yalch and Spangenberg (2000)[70] found that playing familiar (top 40) music in a supermarket led to shorter shopping times, possibly because people found it more stimulating than the unfamiliar music it was compared to.

Texture can also powerfully affect purchasing behaviour. Areni and Kim (1993)[71] found that classical music caused people to spend more money in a wine shop – not because they bought more bottles, but because they chose more expensive wines. Similarly powerful effects from the texture of music were recorded by North, Hargreaves and McKendrick (1997)[72], who found that, in a supermarket displaying both French and German wines with equal emphasis, the playing of French accordion music caused French wine sales to outnumber German by 5:1, while the playing of German oompah music reversed the position, increasing sales of German wine to almost twice those of the French wine.

Loudness, too, can affect a number of behaviours. Smith and Curnow (1966)[73] tested different volumes of music in two supermarkets and found that loud music caused people to spend less time in the store (17.64 minutes compared to 18.53 for quieter music). The people spent the same amount of money in each soundscape, which means they spent faster in the loud music soundscape. Sullivan (2002)[74] found that soft music led to longer meal times in a restaurant compared to either loud music or no music, with a resulting increase in spending. Guéguen, Hélène, and Jacob (2004)[75] found that loud music in a bar caused people to buy more drinks than soft music.

Crossmodal effects

Sound does not ever exist in isolation. Even with our eyes shut or in complete darkness, we imagine visual images when we hear sound. Most of the time we are receiving data from all our senses simultaneously, and it has become clear in recent years that they affect one another significantly. These effects are described in academic circles as crossmodal, and interest in them (both academic and popular) has surged as their power has been revealed.

The UK's leading researcher in the field is Professor Charles Spence at Oxford University, with whom I have had the pleasure of sharing several speaking platforms. Among many other fascinating crossmodal effects involving sound, Spence's research has shown that:

- people associate tones or frequency bands with particular tastes (for example, coffee is associated with low frequencies, presumably because it's dark and rich);[76]
- people's perception of texture can be changed by altering the sound they receive as they are eating (boosting 2 kHz made people think Pringles crisps were crisper);[77]
- people think aerosols are more effective if they sound more powerful.[78]

Spence also worked with well-known multi-Michelin starred chef Heston Blumenthal to create his signature dish at the Fat Duck restaurant. It's called 'Sounds of the Sea' and it requires the diner to wear headphones which deliver a seashore soundscape to enhance the taste of the dish. Spence has focused more recently on the multisensory design of cars, to improve driver awareness and reduce accidents. You can hear a short interview I did with Charles Spence on my AudioBoo page.

Taste is not the only sense that relates to sound. The famous McGirk illusion shows that what we see profoundly affects what we hear.*● It also modulates the way sound affects us: for example, if you see a huge crane ball heading for a wall, your hearing automatically compensates and you are less startled by the ensuing crash – a form of internal audio compression that your brain carries out to keep you from being continually shocked and frightened by sudden but innocuous noises. The same noise on headphones without the visual cues is far more disturbing and powerful.

Brands have been using crossmodal effects for years. Bang & Olufson adds weights to its remote controls to make them feel higher quality. Supermarkets pump the smells from their in-store bakeries back into the stores to create a more attractive ambiance.

---

* ● You can see and hear my version of this illusion on the website.

When the senses are brought into harmony, with congruent messages coming from all of them, what happens is a phenomenon called super-additivity: the combined effect is multiplicative, not additive. The reason it's so important for business to become aware of sound is that the reverse applies: if sound is incongruent, it undermines the effect of visual branding and communication, by over 80% according to Spence.

We have discovered by moving through the whole SoundFlow™ model that sound affects us in many ways, such as entrainment by tempo, arousal by pitch and texture, distraction by density, and emotional effects, especially of music.

Many of these chains of cause and effect act in opposite directions so it is difficult to generalise. When we audit sound in a space or for a brand, we run SoundFlow™ from top to bottom and consider other senses to gather all the individual circumstances and then come to a view about what the sound is doing (or could be doing).

Part 2 References

21 Spaces Speak, Are You Listening? Experiencing Aural Architecture (2009), Barry Blesser, MIT Press, p 69

22 Blessr, ibid. , p 5

23 Charlene Spretnak (1999) The Resurgence of the Real: Body, Nature and Place in a Hypermodern World, Taylor & Francis, p 114

24 Mithen, ibid

25 Philip Ball, The Music Instinct, Bodley Head, 2010, p 253

26 Donald A Hodges (ed) (1996) Handbook of Music Psychology (Second Edition), Institute for Music Research Press, p 59

27 Tomasello, M. (2008). Origins of Human Communication. MIT Press.

28 Hodges, ibid, pp 4 and 17

29 Hodges and Haack, The Influence of Music on Human Behaviour, published in Hodges (ed) ibid.

30 International Federation of the Phonographic Industry press release, London, April 28, 2010

31 David Rothenberg (2006) Why Birds Sing, Penguin, also CD Terra Nova Music, 2005

32 The Observer, March 28th 2004

33 Schafer, ibid, p 186

34 Bart Kosko (2006) Noise, Viking

35 Kosko, ibid, p 28

36 Suter (1991) Noise and its Effects

37 Newsweek, April 5, 2004, p 56.

38 Immel, Richard (1995) Shhh...Those Peculiar People Are Listening. Smithsonian. 26(1):151- 160.

39 San Diego Union-Tribune, November 9, 2005

40 Sound To Make an Army Flee (New Sonic Weapon Uses Baby's Cry). The Scotsman. 24 Jun. 2002

41 See FeONIC's website for details at www.feonic.com

42 M G Rigg (1937) Musical expression: an investigation of the theories of Erich Sorantin, published in Journal of Experimental Psychology 21, pp 223-229

43 K Hevner (1936) Experimental Studies of the Elements of Expression in Music, published in American Journal of Psychology, 48, pp 246-268

44 M G Rigg ibid

45 B M Thompson & S R Andrews (1999) The Emerging Field of Sound Training, published in Engineering in Medicine and Biology 18, no 2, p 92

46 Rauscher, F.H. , Shaw, G.L., & Ky, K.N. (1993) Music and spatial task performance published in Nature, 365, p 611

47 Langer SK Philosophy in a New Key: A Study in the Symbols of Reason, Rite, and Art (1942)

48 Stern D (1985) The interpersonal world of the infant: a view from psychoanalysis and developmental psychology, Basic Books, p 67

49 Herrington, J. D. and Capella, L. M. (1996) Effects of music in service environments: a field

study, Journal of Services Marketing, 10, pp 26-41.

50 Sullivan, M. (2002) The impact of pitch, volume and tempo on the atmospheric effects of music, International Journal of Retail and Distribution Management, 30, 323-330.

51 Strogatz, ibid, p 283. Strogatz also reviews some of the key work on brain activity and cognition/ mental states in the previous pages from p 277.

52 Strogatz, ibid, p 282

53 One example is Jevning, R., Wallace, R. K., & Beidenbach, M. (1992) The physiology of meditation: A review. A wakeful hypnometabolic integrated response published in Neuroscience and Behavioral Reviews, 16, pp 415-424.

54 Smith, J. C., Marsh, J. T., & Brown, W. S. (1975) Far-field recorded frequency-following responses: Evidence for the locus of brainstem sources, published in Electroencephalography and Clinical Neurophysiology, 39, pp 465-472

55 For some further reading and contacts I suggest you visit the website of a fascinating conference at Stanford University called Brainwave Entrainment to External Rhythmic Stimuli: Interdisciplinary Research and Clinical Perspectives (May 2006) at http://sica.stanford.edu/ events/brainwaves/schedule.html

56 Melinda Maxfield Effects of Rhythmic Drumming on EEG and Subjective Experience, from proceedings of "Consciousness at the Edge: Shifting Scientific and Personal Paradigms", the 7th Conference on Treatment and Research of Experienced Anomalous Trauma, April 27 – 30 1995

57 See Joshua Leeds (2001) The Power of Sound, Healing Arts Press, pp178-182 for a summary

58 The Sound of Things to Come, New York Times, 23 March 2003

59 To go deeper into this whole area, try Davis (ed) (1982) Interaction Rhythms: Periodicity in Communication Behaviour, Human Sciences Press, and also Brown & Graeber (eds) (1982) Rhythmic Aspects of Behaviour, Lawrence Erlbaum Associates

60 Fried, R. and Berkowitz L. (1979) Music hath charms ... and can influence helpfulness. Journal of Applied Social Psychology, 9, 199-208.

61 North, A. C., Tarrant, M., and Hargreaves, D. J. (2004) The effects of music on helping behaviour: A field study. Environment and Behavior, 36, 266-275.

62 Dube, L., Chebat, J.-C., and Morin, S. (1995) The effects of background music on consumers' desire to affiliate in buyer-seller interactions. Published in Psychology and Marketing, 12, 305-319.

63 For a review of some of the many papers in this area, see Hodges, ibid, pp 520-522 and Leeds, ibid, pp 107-113

64 Berlyne, D. E. (1971) Aesthetics and psychobiology, Appleton-Century-Crofts

65 North, A. C. and Hargreaves, D. J. (1999) Can music move people? The effects of musical complexity and silence on waiting time. Environment and Behaviour, 31, pp 136-149; also North, A. C., Hargreaves, D. J., and McKendrick, J. (1999) Music and on-hold waiting time. British Journal of Psychology, 90, pp 161-164.

66 Banbury, S. and Berry, D. C. (1998) Disruption of office-related tasks by speech and office noise. British Journal of Psychology, 89, pp 499-517.

67 Milliman, R. E., (1982) Using background music to affect the behavior of supermarket shoppers, published in Journal of Marketing, 46, pp 86-91.

68 Milliman, R. E. (1986) The influence of background music on the behavior of restaurant patrons, published in Journal of Consumer Research, 13, pp 286-289.

69 Roballey, T. C., McGreevy, C., Rongo, R. R., Schwantes, M. L., Steger, P. J., Wininger, M. A., and Gardner, E. B. (1985) The effect of music on eating behavior, published in Bulletin of the Psychonomic Society, 23, pp 221-222 and also McElrea, H. and Standing, L. (1992) Fast music causes fast drinking, published in Perceptual and Motor Skills, 75, p362.

70 Yalch, R. F. and Spangenberg, E. R. (2000) The effects of music in a retail setting on real and perceived shopping times, published in Journal of Business Research, 49, pp 139-147.

71 Areni, C. S. and Kim, D. (1993) The influence of background music on shopping behavior: classical versus top-forty music in a wine store, published in Advances in Consumer Research, 20, 336-340.

72 North, A. C., Hargreaves, D. J., and McKendrick, J. (1997) In-store music affects product choice. Nature, 390, 132.

73 Smith, P. C. and Curnow, R. (1966) 'Arousal hypothesis' and the effects of music on purchasing behavior. Journal of Applied Psychology, 50, 255-256.

74 Sullivan, M. (2002) The impact of pitch, volume and tempo on the atmospheric effects of music. International Journal of Retail and Distribution Management, 30, 323-330.

75 Gueguen, N, Helene, L., and Jacob, C. (2004) Sound level of background music and alcohol consumption: an empirical evaluation. Perceptual and Motor Skills, 99, 34- 38.

76 Crisinel, A. S., & Spence, C. (2010). A sweet sound? Exploring implicit associations between basic tastes and pitch. Perception, 39, 417-425.

77 Crisp sounds An experiment to get your teeth into Marc Abrahams The Guardian, Tuesday 23 May 2006

78 Spence, C., & Zampini, M. (2007). Affective design: Modulating the pleasantness and forcefulness of aerosol sprays by manipulating aerosol spraying sounds. CoDesign, 3, Supplement 1, 109-123.

Part 3

Sound Practice

I like to listen. I have learned a great deal from listening carefully. Most people never listen.
Ernest Hemingway

3.1 Sound and brands

We've seen that all businesses are making sound, and that most of the sound around us in the modern world is made by those businesses; we've also realised that most of them just aren't conscious about the sound they output, or its effects on staff, customers and society. Fortunately for all of us, sounding good is not just a matter of corporate social responsibility: almost any business can increase sales, customer satisfaction and brand value by optimising its sound. As this is becoming more widely known, the mechanism of competition is causing more and more organisations to take control of their sound and design their sonic footprint in every aspect of their operation.

In this third section of the book, we review many of the ways sound can be optimised in practical situations. To do this, we will adopt the perspective that in my experience best shows how sound can be directly linked to profit: that is, the relationship between sound and brands.

Well over a thousand billion dollars are spent worldwide every year on how brands look. Visual identities alone can cost millions; most major brands make sure that their sacred visual identity is consistently implemented by creating a brand book which lists every possible usage, from TV advertising to packaging to letterheads and licensed merchandise, and the associated rules for each one. Never this style with that; never in red; always with this tagline; always in this font... these books can be as thick as telephone directories for the world's most famous and complex brands. They are designed to cover every possible aspect of branding.

So it's strange that I have had the following conversation many times, and it is always the same.

Me: "Do you have a brand book?"
Marketing director: "Yes, of course we have a brand book."
Me: "How many pages are about sound?"
Marketing director: "Er, none."

Sound is probably the last great unexplored country for the marketing profession. It is a largely virgin territory, rich with resources, that's been there from the start; we have just been deaf to its potential, such has been our ocular obsession.

Of course, sound is not the only sense we've been ignoring. The other two primary senses (smell and taste) and the range of touch, or tactile, senses (pressure, texture, temperature, balance and so on) are important too. Using the traditional Aristotelian five-sense model (sight, hearing, smell, touch and taste), marketing guru Martin Lindstrom proposed '5D branding' in his 2005 book *BRANDsense*.[79] His extensive research with Millward Brown found that less than 10 per cent of the world's top brands had a sensory branding platform (though this was forecast to increase to 35 per cent within five years, a process that has clearly been happening).

I fully support the 5D approach, in accordance with a principle that Bull and Black in their book *The Auditory Culture Reader* call 'democracy of the senses'.[80] However, sight and hearing must be considered the twin major senses for two reasons. First, they can both carry specific messages: we can say exactly what we want in either vision or sound. Smell, touch and taste can convey a large number of moods, feelings and ambiences, but not many specific messages. Second, sight and hearing can both be broadcast, and they are therefore the only two mass communication senses. So far, nobody has found a way to broadcast smells or tastes. This is not to belittle the important roles these senses play in complete branding, or what Lindstrom calls creating a Holistic Selling Proposition. There are experts already working in aroma who understand exactly how to enhance a brand with the right ambience through carefully designed and deployed scents. In the UK, The Aroma Company has been doing this for major retail brands for many years. But this book is an exploration of sound, so we'll leave these territories for others to explore.

We know that sight and hearing are the two major senses, and we know also that sound has profound effects on people. From this perspective it's clear that the marketing profession has always been out of balance in the weight it gives to sight compared to hearing. This may be because marketing's thinking has been dominated by the mass communication media, which were sight-only (press and posters) for much of its formative history. It may also be that marketing's whole strategic paradigm has been dominated by advertising, and by the brand as promise; brand experience is a relatively young discipline.

But as we know today, a brand *is* both a promise and an experience. Sound can play a major a role in both these aspects, though how major depends on the specific product, brand, market, territory and customer base. At its most potent, sound can make or break a brand. It must always be considered.

It does seem that this message is starting to be heard. A survey by German sensory branding specialists MetaDesign published in January 2006[81] found that managers of top global and German companies, consultants and journalists all agreed that it is crucial to engage all the senses, and that they increasingly recognise that sound can be used in brand management to achieve differentiation, recognition, emotional appeal and identification. However, it was also clear that the group lacked confidence and methodology in applying sound.

Maybe that's why over 95 per cent of Fortune 500 company websites are still purely visual.

3.2 BrandSound™

At The Sound Agency we define BrandSound™ as "all the intentional sound which communicates and/or reinforces the qualities, tones and values that are best aligned with the brand promise, and which enhances and amplifies the desirable aspects of the brand experience." The key word in there is intentional. In a perfect world every business would generate only BrandSound™, but right now it's almost certain that the vast majority of business sound is unintentional and accidental – and a lot of it is counterproductive, or even downright unpleasant.

All organisations are making sound every moment they operate – but they have no clear inventory of it, nor any defined processes to optimise it, to make it consistent with the rest of their branding, or to map its effects. Even in the rare cases where it is well-conceived and executed, most examples of BrandSound™ are swamped by a cacophony of uncontrolled, unintentional sound that nobody is monitoring – sound which the marketing team is unaware of, because it's not made by them. For example, I have lost count of the times I've been in coffee shops where the carefully chosen music is inaudible over the crashing, hissing, gurgling and conversation in a room with shocking acoustics.

Also, as discussed in the last two sections, we must always frame our consideration of BrandSound™ in the context of a fully multisensory approach. It has been fascinating working closely with the sensory experts at BRANDsense agency over the last few years for clients all over the world and discovering how our work can harmonise to achieve super additivity in practice.

The power of this kind of full-scale sensory branding is now widely appreciated: there is a growing literature on the subject in the wake of Martin Lindstrom's work, such the Swedish book *Sensory Marketing*[82] – and many top brands and their CMOs are now enthusiastically embracing sensory branding because they understand that it can achieve an emotional effect that transcends conscious, negotiated acceptance. According to psychologist and branding guru Robert Passikoff, the consumer decision process is 70 per cent emotional and only 30 per cent rational.[83] The senses, especially hearing and smell, are the doorways to our emotions.

Case study: Helm Bank

Tools: workshops, BrandSound™ audit and guidelines, brand music, sonic logo, telephone sound, brand voice, soundscapes, branded audio

Bogota-based Banco de Crédito boldly broke the traditionalist mould of Colombian banking by deciding to rebrand in all five senses as it moved into retail banking. It changed its name to Helm Bank, radically redesigned its corporate visual identity, and deployed sound, scent and taste to deliver the world's first multisensory brand=k brand experience.

BRAND sense agency led the project, and engaged The Sound Agency to overhaul the bank's sound. High-level workshops and interviews and a BrandSound™ audit led to a complete set of BrandSound™ guidelines covering all eight expressions of the brand in sound. Implementation followed, including:

- *brand music in the form of a 'brand suite';*
- *a sonic logo that was so catchy that most staff installed it on their personal mobiles;*
- *two generative soundscapes for the branches;*
- *delivery of a further rainforest soundscape through hypersonic loudspeakers;*
- *consultancy on IVR telephone system design;*
- *choice of appropriate voices for recorded announcements;*
- *on-brand music playlists for office spaces;*
- *all the sonic elements to create consistent high quality podcasts locally.*

You can hear a sample of the Helm soundscape, and the sonic logo, on the website.

●

BrandSound™ Guidelines

As a starting point, every brand should have what we call BrandSound™ Guidelines (BGs) in place as part of its brand book. BGs will list all the practical rules that define the brand in sound. What is the sound of this brand? What is its rhythm, its tempo? Which melodies, harmonies, modes does it prefer, and which does it avoid? Which instruments or specific sounds are on-brand, and which off limits? What is the voice of the brand – what exactly does it sound like? How do we make this consistent in all our voice communication, including the telephone? Does the brand have a particular sound that defines it, or even a sonic logo? If so, how and where is this to be deployed? What is unchanging and non-negotiable in our briefs to composers and sound designers, or our commissioning of commercial music for advertising? What sound is consistent in all our advertising (and maybe also events, promotions, sponsorships)? What soundscapes do we install in each of our spaces, delivered with what equipment or quality specifications, and how do we maintain their quality, freshness and integrity? Is our product sound a differentiating factor, in which case what are the ways in which we nourish and protect that asset?

For some, BGs will be a single page. For other, more complex or major brands, there may be many pages, plus audio samples and examples. However it works individually, one thing is true for all: no brand should be facing the 21st century without a clear statement of its sound.

The eight expressions of a brand in sound

Our BrandSound™ model (below) identifies eight expressions of a brand in sound, with BGs at their centre. There are other models around and I recommend you try them to see which one best suits you – the Audio Branding Academy's yearbooks[84] are a good place to start. This one has worked well for us over the last five years and so far we haven't come across a client for whom it hasn't delivered great clarity and value.

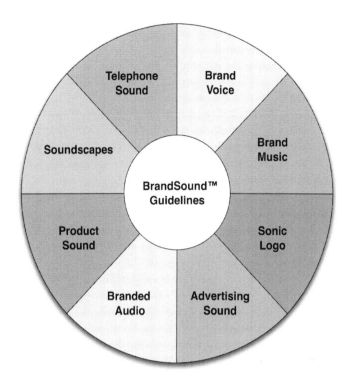

- **Brand voice** is not a metaphor: it is how the brand sounds when you hear it speaking through any one of thousands of vocal interactions, from the receptionist or operator who greets you to the professional voice over artists it uses to the sales person who pitches to you.
- **Brand music** is any music that's strongly associated with the brand, be it licensed commercial music or specially composed brand themes of songs.
- **Sonic logo** is a short (usually three to five second) sonic mnemonic – the aural equivalent of the visual logo.
- **Advertising sound** is all the sound involved in TV or cinema commercials, comprising voice, music and sound effects, with the whole definitely greater than the sum of the parts.

- **Branded audio** is sound created with the brand imprimatur on it. Formerly this would have been exemplified by branded CD compilations but now it means podcasts, audio downloads, audio books, webinars, teleseminars and more.
- **Product sound** is the sound of using the product or service. In some cases this is non-existent, or covered in one of the other categories; in others it is intrinsic to the product's value.
- **Soundscapes** describe either the totality of sound in a commercial space like a shop or a corporate reception, or more specifically a designed stream of sound intended to create a particular effect on the people in such a space.
- **Telephone sound** means all sound in telephone interactions, from ring tones and on-hold sound to recorded announcements, automated call-handling systems and call centre agents.

In the following sections we discuss each of these expressions in detail, including an in-depth and very practical review of soundscapes in many different types of commercial space (both physical and virtual), and of telephone sound.

In practice, before we get into writing BrandSound™ guidelines and creating (or removing) actual sound, there are two important steps to take.

First is a proper audit. We have done this for shopping centres and major department stores, for airports, and for whole brands. It's essential to find out what's so and where we are, before planning what will be and where we're going.

Second is a workshop for the company's senior management. Most of them will never have thought about sound, and in order to gain their support and engagement with the process of optimising the organisation's sound, it's vital to sensitise them and achieve buy-in. We run a powerful and highly visceral workshop where we demonstrate how important sound is, present the results of the audit, and do a set of intensive interactive exercises that lay the foundations for the work that's to come.*

Those steps having been taken, we can move on to consider each type of BrandSound™ in turn.

---

* A do-it-yourself version of the whole workshop, including the exercises, is on my BrandSound™ DVD. For details visit www.juliantreasure.com.

Case study: Siemens AG Worldwide

Tools: brand music, sonic logo, soundscapes, brand voice

Siemens has a long history of pioneering sound: in 1950 it commissioned the composer Riedl to create ground breaking electronic music for its corporate film Impuls unserer Zeit, and in the 1960s it sponsored a studio in which Ligeti, Cage, Stockhausen and others made music.

In 2003 the company commissioned Berlin-based agency MetaDesign to create a palette of musical sounds, including a sonic logo, to express its brand personality (excellent, innovative, responsible). The agency used the golden section, in the form of a Fibonacci algorithm, to express 'the pulse of life – the impulse for innovation'.

The palette includes corporate music, a brand soundscapes, a corporate voice, and an audio signature (sonic logo). The basic building block for musical elements is the Siemens motif:

Siemens Motif
Composed by Rainer Kirchmann, Jan Michael Pieper, Joerg Christian Hartje, Carl Ton – Published by MetaDesign AG (GEMA 117138), 2007

There are soundscapes for many uses (for example shows, films, on-hold); these employ natural sound as well as designed sound, and can be combined and layered to give exactly the right effect

The company distributes these assets globally via a brand management intranet called /brandville to ensure a consistent sound wherever it operates – thought the platform also keep all the sound assets evolving as people around the world contribute new treatments.

You can hear the Siemens brand palettes on the website. ●

3.3 Brand voice

In the section on the human voice, we reflected on its power, its attributes and the ways in which it affects listeners. Every brand that expresses itself through voice needs to define its default brand voice in order to become consistent in the way it is heard. The importance of brand voice varies of course, but when you consider that brands such as modern banks and utilities are experienced almost entirely in sound (in the form of telephone conversations), it is astonishing to realise that they pay little or no attention to the way people speak. They may train in what to say, but almost never in how to say it. As a result, successive phone calls can produce wildly inconsistent experiences of the brand – one minute jokey and informal, the next reserved and morose, and a third time professional and brisk.

A study by Soukup showed that 'speech is more powerful than the written word or a visual image in making a good impression'.[85] In order to make a good and consistent impression, we define the default brand voice by asking a series of questions and by seeking the answers at the core – that is, in the brand values and the brand personality, and possibly in the corporate culture. Standing firmly in the centre of the brand we ask: is its voice masculine or feminine? Is it mature or youthful? Urgent or laid-back? What register doe it use, and what timbre? Does it have an accent or a particular vocabulary? Defining the voice in these terms helps in creating consistent advertising, and in training anyone who are to represent the brand with their voice. The default brand voice will also inform the choices of telephone operators, announcers, recorded voice announcements, sales people and all other voice contact points.

Of course the guidelines should be so tightly drawn that everyone has to speak in exactly the same way: we are not seeking to create an army of robots. However every brand should be conscious of what its default voice is, and what is acceptable deviation from it,. Some companies do have an implicit understanding of these things, but the ways they are transmitted are more often than not myth, story, working practice or hearsay ("Did you hear? Fred got sacked for saying 'That's crap' to a customer"). The brand voice guidelines must be clear, consistent, reliable and well communicated to everyone, so that brand experience at grass roots level actually *is* what the management fondly imagine it to be.

Case study: Companhia de Saneamento Básico do Estado de São Paulo S.A (Sabesp)

Tools: brand music, sonic logo, brand voice, telephone sound

Sabesp is Brazil's largest water company. Its management felt the visual brand needed refreshment, and decided to use sound to achieve this. They retained São Paulo based agency Zanna Sound, who used a six-step approach:

1   *Brand study: reviewed history, research and current branding to define their personality and archetype.*

2   *Sound communication audit: reviewed the client's sound communication and that of its competition; researched 524 people about the client's brand sound. The research concluded that the brand sound should be feminine, mature, and use eclectic/Brazilian genres of music.*

3   *Sound guidelines for: musical theme; sonic logo; brand voice; sound design; repertoire.*

4   *Creative and production for all the above core sounds.*

5   *Development for telephone, web site, radio ads, podcast and films.*

6   *Two workshops to increase awareness of sound branding communication.*

You can see a short film and hear the brand music and sonic logo on the website.

●

3.4 Brand music

Some brands have won great appeal by associating themselves with a piece of music. One good example of this is British Airways' use of the Flower Duet (*Sous le dôme épais*) from Léo Delibes' opera *Lakmé*. BA doesn't own this piece of music: it has been used in many films, from *Meet The Parents* to *Carlito's Way*. But no other brand would be able to use this music now without either parodying or in some other way making reference to BA, such is the association in the minds of the TV viewing public, in the UK at least. BA has also used the music as its on-hold sound, and on its fleet as background music while boarding and after landing. The music communicates peace, calm, beauty, class, distinction, refinement – all in all, an inspired choice and one which continues to benefit BA today.

This example shows how to form a powerful and positive association by creating brand music from a pre-existing commercial piece at a modest cost. Even if the music is far more expensive, as is becoming common in the vibrant 'band/brand' market, where brands are effectively becoming the modern-day patrons of the crippled music industry, the effects can be well worth it. According to North and Hargreaves, consumers are "24 per cent more likely to buy a product with music that they recall, like and fits the brand compared with eight per cent when the opposite applies".[86]

More recently, brands like Lufthansa and Siemens have begun to create their own tailor-made music. (For more on Lufthansa and Siemens, see the following section on sonic logos.) In the future, I hope we will see brands applying some of the principles of psychoacoustics described in this book to make sure that the music they create and identify with (or make part of) their brand is actually appropriate, effective and consistent with the rest of their branding.

Brand music can include promotions or alliances such as the 2005 relationship between T-Mobile and Robbie Williams, where the artist's tracks were made available early or even exclusively to T-Mobile subscribers. However, this kind of deal is more in the domain of celebrity endorsement and does not identify the brand with a particular sound to the same degree. Brand music is most powerful when it is the identifiable musical expression of the brand identity, not just a piece of mutually beneficial joint promotion.

3.5 Sonic logo

Visual logos are not brands, of course, but each is intended to represent the essence of its brand's character – to introduce it if we don't know it, or to remind us of it if we do. As a photograph is to a person, a logo is to a brand. Ogilvy's Rory Sutherland likes to define a brand as a piece of data compression, and the logo plays an important role in capturing and conveying much of this data in an incredibly small space.

We encounter thousands of logos every day, but almost all of them are visual. For the next few years, sonic logos will always be worth considering simply because they will be relatively rare and can therefore act as differentiators. But there's more to them than curiosity value: used wisely, they work exceptionally well.

Sonic logos have actually been around for hundreds of years: street calling used to be the main way tradesmen advertised their services, as romanticised in the film *Oliver*. It's not so long since that practice died: I can remember the rag-and-bone man's mournful shout of "Anyoldiron?" from my childhood. The modern-day equivalent is the ice cream van: just watch the cathartic effect of its chimes on surrounding buildings on a hot Summer's day to see the potency of sonic logos deployed in the right place at the right time. In the UK ice cream chimes are fairly generic, but in Sweden the Hemglass ice cream tune is a universally known and loved manifestation of an individual brand. *●

As soon as sound recording became viable, the advertising industry saw the potential of memorable music/voice combinations and the jingle and tagline were born. The dividing line between jingle or a tagline and a sonic logo is blurred. In general, jingles and taglines come and go with campaigns, or are specific to them, and rarely live for more than a few years. Even the most memorable usually get retired. "For hands that do dishes…"; "1001 cleans a big, big carpet…"; "*Vorsrpung durch Technik*, as they say in Germany"… once mighty, all now languishing

---

* ● The Hemglass ditty is on the website, along with the Intel logo, the Mac switch-on sound and the MGM lion.

in their bath chairs, though two of the three brands are still very much with us today. Some taglines are so strong they become sonic logos and one in particular has outlasted entire generations of customers: Tony the tiger has been saying "they're gr-r-r-r-reat!" since 1951. This is probably the longest-running sonic logo in the world, and it has now outlived its voice-over artist. Thurl Ravenscroft was famous for many Disney voices but Tony was his greatest legacy. He voiced the tiger for 54 years until his death in 2005, when Lee Marshall was appointed to carry the tradition forward.

Tony notwithstanding, it wasn't until the 1990s that sonic logos started to be taken seriously and their use came to be considered by major brands as a matter of course. In London, Dan Jackson set up the Sonicbrand agency and wrote his book on the subject[87], but the real sea change came with Intel. Its five-note sonic logo, composed by Austrian musician Walter Werzowa, has become one of the best-known sounds in the world, and has spearheaded Intel's extraordinary success as a brand – given that this is a product nobody ever sees and nobody ever buys.

Today, sonic brands are more in play than ever before. UK insurance giant Direct Line has its bugle call, which speaks volumes about urgency, assistance and playfulness in just three seconds.

Nokia has its ring tone, adapted from Francisco Tárrega's *Gran Vales*. First used in a 1992 ad, the ring tone is now the world's most-played tune, issuing from a mobile device somewhere in the world 20,000 times every second (that's 1.8 billion times a day!) simply because so many people don't bother to change their default ring tone. It was a stroke of genius to use human inertia to leverage viral advertising like this. It remains a mystery why Vodafone, Sony Ericsson and the rest have never challenged Nokia's dominance. Nokia are keenly aware of the need to keep such a commonly heard sound fresh so that it doesn't become irritating, and they constantly produce new arrangements to suit the different characters of various handsets, as well as improving quality as handset audio technology continues to evolve.

Apple has its comforting, uplifting start-up sound, engineered in 1991 by Jim Reekes and still shipping today. It is inexplicable that the mighty Microsoft has never seen the value of a single start-up sound; the sound of Windows has changed with every successive version of the software, so that now there is no sound of Windows. They may be learning though: huge amounts of time and money were invested in 'a language of sounds' for the Xbox 360.

Lufthansa created a corporate sound, comprising four rising tones that are aimed to convey feelings of taking off and wellbeing. Siemens added a seventh element to its branding: sound has now joined logo, claim, typeface, colours, layout and style as one of the basic building blocks of the Siemens brand. The company created both an 'audio signature' (alias sonic logo) and also some mood sound as part of its brand palette. Over the years, some sonic logos have even been registered as trademarks or service marks: the roar of the MGM lion and the old NBC three-tone chime are two examples.

More and more major brands are creating a sonic logo as a matter of course. With the rise of mobile devices (and with them the tailoring of ring tones and the prevalence of downloaded digital sound), I believe we have only scratched the surface of the sonic logo so far. Their efficacy extends beyond the fact that they are not visual, and therefore avoid the overmessaging that makes it so hard for a visual logo to stand out. Sound is at once more intimate that visual input (it actually touches us inside our heads as the sound waves vibrate our eardrums) and at the same time a powerful and primal trigger for emotion, as discussed in the previous section when we reviewed music and sound dynamics. This is why a well-designed sonic logo can be such a potent embodiment and communicator of a brand's personality, and forge such strong emotional connections with customers.

There is little research on the effectiveness (or in the technical language of sound research the affect) of sonic logos. The most comprehensive analysis I have come across is by Copenhagen-based Delta SenseLab's Jesper Ramsgaard. He wrote his 110-page Master's thesis on it[*]. The work had two goals: first to find the best tool for measuring the impact of sonic logos, and second to take a first step in indicating what it might be using a sample of sonic logos produced by Danish agency Sonic Branding and a small group of consumers.

Ramsgaard tested three approaches. The first was simple preference, which proved a blunt instrument for assessing any shades of emotion.

---

[*]   Delta have kindly issued a précis; either the full thesis or the summary are available from the author at jra@delta.dk.

The second was Swedish Core Affect Scales (SACA), which is an application of the wider theory known as circumplex. This suggests that all human emotion can be placed in a space with two dimensions: arousal (from passive/dull to active/lively) and valence (from sad/displeasing to glad/pleasing). A simple circumplex model is shown below. The axes are often transposed in the research literature, with valence horizontal and arousal vertical.

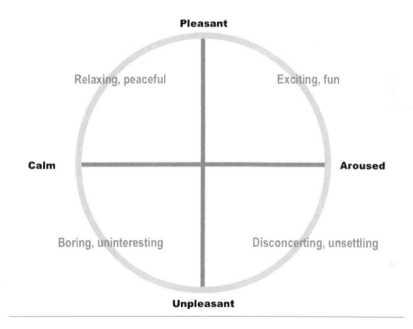

SACA proved much more refined than simple preference, and there was plenty of differentiation between the ten sounds tested. However, valence is again very open to interpretation, which created noise and implied that a larger sample with some sort of sub grouping would be needed to get a clean signal.

Finally, Ramsgaard tried Zentner et al's Geneva Emotional Musical Scale (GEMS)[88], which scores sound on nine emotional scales: tenderness, power, wonder, sadness, nostalgia, joyful activation, tension, peacefulness and transcendence. Despite a smaller sample of people, the result were significant and furthermore it was obvious that the respondents instinctively understood how to use the tool, making their ratings much more accurate and the results probably very repeatable.

The GEMS tool clearly showed the very different affective profiles of the sonic logos: two of the results are shown below, reproduced here with acknowledgement from Ramsgaard's work. The circles around the data points indicate 95 per cent confidence intervals.

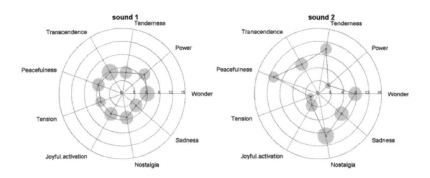

Fascinatingly for those of us who make sonic logos, the GEMS method offers us a powerful and consistent briefing tool. It should be possible to analyse the brand personality in these nine dimensions, plot it on a GEMS graph, and then use that to brief a composer or sound designer – and to measure the prospective sonic logos by testing them on sample groups to see if they deliver the right emotional impact. With this kind of structured approach we can move on from the sound branding specialist's least favourite conversation, where a client calls up just wanting a 'bing-bong'. Tools like GEMS can at last help us to design and deliver accurate sound – and to test it and prove that it works.

Case study: LBS, Germany

Tools: sound audit, sonic logo

With a market share of 33% and nine million customers, the LBS group is Germany's number one building society in a fiercely competitive market. Research showed the LBS brand was losing ground to its main competitor, so the company decided to use the power of sound to regain its primacy. It briefed Hamburg-based GROVES Sound Communications to audit its current use of sound.

Using its well-developed Sound Branding methodology, GROVES discovered that the problem was not just sound but also cluttered and unclear TV endings. Working with BBDO Berlin, a new clearer treatment was developed by commissioning a visual logo animation, adapting the existing LBS music asset into a well-defined sonic logo, and using production company Spann & Partner to integrate the two.

The cleanup worked. In focus groups, independent research institute Emnid found that 76% of the participants correctly identified the sound logo from a choice of brands; 61% could spontaneously name the brand; and 86% recognised the melody. The new sound identity was shown to significantly improve recall, both with and without visuals.

Sound had put the LBS brand back ahead of its competitors.

3.6 Advertising sound

One form of advertising is pure sound, of course. Radio advertising has been around for many years, so it is clearly effective. Research by Radio Ad Effectiveness Lab[89] produced results that have been found before: people form a deeper emotional bond with radio than with press or TV, and they find advertising on the radio more personally relevant. This is probably because of the more passive, intimate nature of hearing compared to seeing. When we read a paper, or watch TV, we are doing something (looking) and we can't really do much else. When we listen to the radio we simply allow it in, and we very often are doing something else, like driving or ironing, or (according to recent research) browsing the Internet, which has forged a potent link between hearing and buying.

At its best, radio becomes part of us, not part of the outside world. Also, we generally hear one voice on the radio, and we happily and knowingly subscribe to the convention that that voice is talking only to us. Radio is seen by many people as a companion in a way that other media rarely are. In this context it's no surprise that radio advertising is enduringly effective.

There have been several academic studies on the way music affects responses to TV or cinema advertising (though much less research has been done into the other aspects of sound, such as the voiceover, sound effects, mix and volume).

It is clear from the public domain research that music in advertising can have a significant effect, increasing recall and producing positive or negative associations. This comes as no surprise: if agency internal research hadn't been showing this for years, then advertisements would have long since been music-free. A typical finding from the academic world is that of Alpert and Alpert in 1989:

"Variations in the formal music structure of background music in commercials may have significant influence over the emotional responses of an audience."[90]

In tune with this result, Park and Young (1986)[91] found that music can enhance attitudes to brands. Anyone who remembers the long-running Hamlet cigar TV advertising in the UK can testify to the strength of the association between Jacques Loussier's cool version of Bach's *Air on a G String* and the Hamlet brand, which developed overtones of coolness, class and wry humour as a result – not bad for a low-cost, mass-produced cigar.

Morris and Boone (1998) found that:

> "Music may not always significantly change pleasure, arousal, dominance, brand attitude, or purchase intent in an emotional advertising condition, but it can change how the viewer feels when watching the advertisement."[92]

It is still a matter of debate whether music achieves increased advertising effectiveness by creating positive associations, or by actually changing the viewer's values about the product, or both.

However it works, well-chosen advertising music, like film music, can send a shiver up the spine or make a commercial eternally memorable. The multi-award-winning Guinness *White Horses* ad would surely never have been the experience it was without the inspired choice of the opening section of Leftfield's *Phat Planet* as its main sound. Coca-Cola's *Real Thing* ads in 1970 were not the first, and will not be the last, to spawn a hit single or make a career in the music business. A recent survey by McCann-Erickson in the UK revealed that the Wall's Cornetto reworking of the Neapolitan folk song *O sole mio* (*Just one Cornetto*) was the best-remembered advertising jingle of all time: 70 per cent of those surveyed still remembered it, 24 years after the campaign finished. The tune was re-enlisted in a new commercial as a result.

Advertising sound might be called the frivolous younger sibling of film sound, and it certainly borrows many of its strengths and techniques from that source. The film industry is the most experienced in the world at using sound for emotional effect; it has done so since the beginning of sound in cinemas. In a wonderful article called *Sound Tricks of Mickey Mouse*, which appeared in a publication called *Inventions* way back in in 1937[93], Earl Theisen writes:

> "Through study and experimentation Walt Disney and his engineers have found that by introducing music or various sounds and noise frequencies into the cartoon, the response of the audience is varied and controlled. By combining noises of certain pitches or tempos the psychological values of

the cartoon music is emphasized in keeping with the story requirements. Sound of sixteen cycles is deep toned and may be used for conveying heavy or depressing moods, whereas the sound of a higher frequency is what William Garity, Chief Engineer at Walt Disney's, calls the 'pain sensitive region.' Noises of this higher pitch make the hearer alert and may be carried to the point of actually causing distress, such as a 'file on glass' noise. The average ear is very sensitive to sounds of 2,000 or 3,000 cycles and unless some sound of this pitch is added to the cartoon background noise, the audience is less responsive to the effects."

There is plenty of good analysis of the science and art of film sound over its long history. French composer and author of several books on the subject Michel Chion concludes in his excellent book *Audio Vision*:

"Sound, much more than the image, can become an insidious means of affective and semantic manipulation. On one hand, sound works on us directly, physiologically (breathing noises in a film can directly affect our own respiration). On the other, sound has an influence on perception: through the phenomenon of added value, it interprets the meaning of the image, and makes us see in the image what we would not otherwise see, or would see differently. And so we see that sound is not at all invested and localized in the same way as the image."[94]

All too often, however, music is an afterthought in the production of TV advertising. A commercial can cost six figures to make and take months to plan and shoot, and then music is chosen in a rush right at the end. I have many composer friends who have suffered at the wrong end of this process: the call comes in asking for a complete piece by tomorrow, with a brief from hell along the lines of "make it like the theme from Rocky but sensitive" and, if they are really unlucky, a session with eight or nine people from client and agency sitting behind them as they work, chipping in with random and conflicting requests based on their personal musical tastes.

If every brand has its BrandSound™ guidelines (BGs) in place, this will not need to happen. Music for advertising can be made consistent (which does not mean all the same) so that, over time, each brand is expressing itself more and more powerfully through the music it commissions or chooses.

That still leaves the problem of finding the right track among millions available. Technology is of limited help: the digitisation of music, combined with modern database technology using tags, is only now starting to crack the problem. There are several websites offering music lovers an automated service that suggests or plays music they are likely to like based on their preferences, either entered by them or imputed from what they've been listening to: Last.fm and Pandora are two interesting examples. There are also plenty of very smart people working in more sophisticated algorithms based on the actual characteristics of each piece of music, so this is going to be a lively area to watch. It is as yet far from the finished article due to the complexity of music – not to mention the fact that nobody knows how it really works.

For commercial applications, London-based Ricall uses technology in a more basic way: its database of three million tracks is searchable by tags for mood, activity, demographic, tempo, lyric, instrument, chart and 'soundslike', and when the track is chosen Ricall also automates the complicated and extensive paperwork involved in securing all the licenses required to use it.

A more refined solution may be to use human experience and human ears, which is how Ruth Simmons of soundlounge and her team work. As noted above, music is notoriously difficult to analyse digitally. Sometimes choosing the perfect track, whether it's for a party compilation or a multi-million dollar TV commercial, requires that flash of insight that can only come from an experienced and knowledgeable human mind; in addition, the value of contacts and reputation can be very high when negotiating rights contracts person to person.

Along with the right music, any voice art that's used must be the brand personified. Choosing voice-over styles campaign by campaign should be resisted unless there are good reasons to change. In the BGs there will be a clear definition on the voice of the brand in terms of age, gender, accent, style, tone, timbre and so on, and these parameters should be adhered to in order to build consistency. Without consistency, where is the brand? It's very rare for any brand to change its visual logo or strap line from campaign to campaign, for this exact reason. Why should sound be any different?

By starting with BGs and then considering the needs of each individual campaign, we can at last ensure that sound will take its proper place as a major contributor to the effectiveness of any TV or cinema campaign, at the same time adding to the power and familiarity of the brand.

Case study: United Nations annual climate conference (UN/COP15, 2009)

Tools: sonic logo (including ringtone), brand music

The annual UN climate conference in Copenhagen was a global event, attended by the world's heads of state of the world, climate organisations, media and the public.

The UN asked Copenhagen-based agency Sonic Minds to develop a sound identity for COP15 that would work in both electronic media (such as the webcast and official YouTube channel) and physical (inside the conference centre and across contact points in Copenhagen and elsewhere).

The agency's solution included brand music, a sonic logo, ringtones, sound design and sound effects. The creative theme combined the concept of climate as a fragile organic amalgamation of micro-processes, with that of Denmark as an open minded, hospitable and climate-concerned host for the conference. The concrete theme was movement from disorder to balance – or in sound terms a transition from many (disagreeing) voices to one (united) voice.

You can hear the COP15 sound work on the website. ●

3.7 Branded Audio

Packaging up valuable sound, putting your brand on the package and selling it or giving it to customers with your compliments is nothing new. Musical compilations have been around for many years, and became big business in the 1980s with the rise of celebrity DJs and the huge success of compilations such as those released under the brands of Buddha Bar, Café del Mar and Hôtel Costes. Each of these organisations has probably made far more money from the music compilations bearing its name than from its core business as a bar or hotel.

Profitable as these compilations could be, there was a substantial entry cost to becoming your own record company, and so creating your own branded album entailed a large risk. If it bombed, the sponsor could lose a lot of money. These days the CD is dying and digital content is opening up a whole new vista for branded audio. We have come to call it podcasting.

The joys of podcasting in sound are that it is relatively cheap to do, even at a very high standard of production, completely scalable at zero (or at worst very low) marginal cost, and highly effective in getting a message across.

At entry level, AudioBoo has made instant, zero-cost micro-podcasting possible. So far it has been mainly a rich form of tweeting, where people follow one another and semi-randomly encounter short audio posts from those they follow. The real potential of this medium is in two other uses, and in neither have we more than scratched the surface so far. First, as a cloud-based resource, in the way that the British Library is constructing its fascinating UK SoundMap, which at the time of writing contained over 1,400 recordings made by unpaid contributors all over the country*. Second, as a branding tool, where a brand can create a programme of 'boos' that customers or fans can subscribe to – for example, a specialist retailer producing a series of interviews with expert staff on topics they really know about; a financial services

---

*   See and hear the map at http://sounds.bl.uk/uksoundmap/index.aspx

company producing a daily boo with news and advice for its clients; or a major consumer brand sponsoring a series of fictional boos like five-minute faux-reality playlets that catch the public imagination.

Moving up the scale from AudioBoo we have an explosion of podcasts, mainly distributed through the iTunes platform (though they can very easily also be disseminated through websites). Alongside the music, the comedy, the celebrities and the news and magazine publishers we are seeing more and more brands producing podcasts.

One good example is Mercedes-Benz, which has been producing its Mixed Tape podcast every eight weeks since 2004; each edition presents listeners with a new group of breaking artists chosen by a panel of music experts. This is a good enhancement of the quality sound systems installed in the cars, and also creates overtones of being leading edge, in the know, cool and trend setting.

Sport is a natural for this kind of activity. Nike produces a series called *The Loop*, interviewing top runners to produce hints and tips for Nike customers. Not to be outdone, Adidas has its own series on skateboarding.

In the beverage/entertainment arena, Bacardi started a series called B-Live, with news and reactions from its B-Live parties – but stopped after just two podcasts. This is a cardinal sin for branded audio. Better not to start at all than to dabble and tease people, or to start and stop. A well researched and targeted podcast series should produce quick and measurable results, and such a programme should be entered into with commitment, vision and the budget to carry on until the job is done. When that happens is a judgment call – but even after the curtain is brought down on a series, one of the joys of podcasting is that the output can last a long time if it's not too date-stamped.

Back in 2005, PR blogger Kevin Duggan posted 20 suggestions for effective branded podcasts. The list is as valid today as it was then. With thanks to Mr Duggan, I have adapted it with additional or updated commentary. Here it is.

1. **Airlines**: travel and destination guides, delivered via the inflight entertainment system and downloadable to mobile devices from your seat; also branded music selections fro the IFE system.
2. **Apparel / Retail**: behind the scenes at fashion shows, interviews

with models and designers, guide to the season, tips and tricks from experts.

3. **Automotive**: test drives of new models with team drivers, interviews with designers, race reports from behind the scenes in NASCAR, F1, rally sport.

4. **Drinks**: Recorded tastings and reviews of selected wines, spirits and beers, interviews with growers and makers.

5. **Books / Music**: guides to this month's (or seasonal) releases, interviews with bands and authors, sample readings from authors to promote personal appearances, diaries of the making of albums or books.

6. **Financial services**: retirement guides, investing tips, planning for school or college education fees.

7. **Golf**: guides to courses, interviews with or diaries from sponsored tour golfers, reviews of equipment.

8. **Grocery**: interviews and recipes from well-known chefs, nutritionists, price news.

9. **Health insurance**: wellness tips, program suggestions, meditation guides, nutrition guides.

10. **Home improvement**: step-by-step guides to every aspect of DIY / home improvement.

11. **Hotels**: destination guides, jogging routes, music from the hotel or music to jog to, walking tours of cities.

12. **Microbrews or destination pubs**: interviews with the brew master or landlord.

13. **Movie studios**: interviews with actors or directors, movie making production diaries.

14. **Sportswear**: exclusive interviews with sponsored athletes, event diaries (Olympics, World Cup), and interviews with designers, fitness and training tips

15. **Travel and tourism**: guides to destinations, travel tips, hotel walkthrough reviews and guides, latest deals and offers with snippet reviews.

16. **Vineyards**: Interviews with wine-makers, vintage notes.

17. **Whisky**: Interviews with the makers, tours of the distilleries, background on water, barrels and other key elements of the product.

18. **Video games**: Interviews with game designers, production diaries, tips and tricks, cheats, panel discussions with expert players.
19. **Higher Education**: communicating with incoming students and providing them with lectures.
20. **Investor relations**: CEO's review, audio annual report with interviews, earnings reports with press conference-style interviews with management and commentators.

This is far from exhaustive: I have no doubt that the list is almost infinite and that you have probably thought of several more while reading it – but it does show something of the breadth and depth of the opportunity for brands to engage with the low-cost, highly effective form of BrandSound™.

I believe branded audio is set to boom in the next few years, with many brands becoming de facto Internet radio stations, attracting relevant and appropriate content around them and delivering it to their customers or prospects in order to add value to the brand, and to express it in new and appealing ways. With HTML5 now allowing sound to be woven into the fabric of websites so that clunky pop-up players are no longer necessary, podcasting in all its forms is a vibrant, rich and exciting medium for all brands to explore, from the largest global players to a corner shop.

3.8 Product sound

Another vital component of BrandSound™ is product sound – the sound of the product or service itself. In some cases, product sound is integral to BrandSound™ and is carefully nourished and protected; in many others it is completely different from its BrandSound™ because it has never been recognised as an asset (or a liability) by the marketing team.

Some products *are* sound, for example a music download or a ring tone. For them, the sound of the product is all there is. Close on their heels are products whose primary function is to make sound, like musical instruments, stereo systems or MP3 players, where sound quality usually outweighs all other characteristics when the product is being considered. At the other end of the spectrum, there are completely mute products, like a book or a pair of trousers, where there is simply no potential for leveraging product sound. (Even these two examples will not be inviolable: I have no doubt that somewhere in the world there are both books and trousers that have a differentiable sound!)

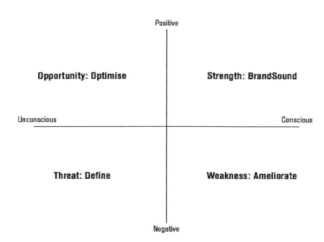

In between these extremes we have a fascinating spectrum of products and services that have a precious asset (or a real liability) in their product sound. Many of them are simply not conscious of it. This creates a range of situations, summarised in the graphic below.

Where your product has positive sound and you know it, this forms an important part of your BrandSound™. Where you have positive sound and you are unconscious of it, there is a major opportunity and you need to scope it and seize it. Where the sound is negative and you know about it, your efforts will be directed at amelioration. And where there is negative product sound of which you are blissfully unaware, you are facing a threat and you need to assess it and deal with it. A perfectly sound-sensitive organisation would have no unconscious sound at all, but in most cases this is an unattainable ideal.

Let's consider some examples of product sound as both an asset and a liability.

Product sound as an asset

We've already discussed products that have managed to incorporate a sonic logo into their operation, like an Apple Mac or a Nokia phone, but these are not quite product sound – the essential sound of the product being used. The world's leading explorer of this interesting territory is undoubtedly the automotive industry, which has long realised that fundamental product sound is often a major selling point or differentiator. If you go to the Ferrari website you can listen to the sounds of the engines of all their models – because it matters how a performance car sounds, just as much as how it looks or performs.

Harley-Davidson went one step further by trying to defend its classic three-beat 'potato-potato-potato' engine sound in court when they perceived that Japanese bike makers were copying it. The case was eventually abandoned after several years, mainly because Harley-Davidson could not prove that their sound was not purely functional: any V-Twin bike is going to make that sound. You can't trademark a functionally necessary feature – that's the realm of patents – so in order to be a trademark a sound needs to be discretionary, even unnecessary – as in the example of the MGM lion we have already mentioned.

Trademarks notwithstanding, there is fierce competition in the automotive industry in sound, and not just regarding engine noise. The level of road noise in the cabin is a major selling point, so car-makers have become experts in acoustic insulation (though even they are outdone here by plane makers, who really are at the leading edge in this field). The sound of doors is another prime battleground. Car doors *au naturel* sound tinny when they close – they are essentially just hollow lumps of metal after all. That satisfying clunk we all enjoy is a completely artificial sound, engineered with padding, weights and rubber to conform to a standard we have all tacitly agreed on.

British car-maker Rover used Danish sound lab Delta to optimise the sound of its power steering. By recording the sound and playing it back through headphones to a panel, Delta identified that attenuating sound in the 280—580 Hz frequency band would make the sound most palatable; Rover changed the length and stiffness of its hydraulic hoses to achieve this effect.

Even car-makers have deaf spots, however. I don't believe that any sound designers or acoustic engineers have been involved in creating the warning sounds that most cars use for everything from seatbelts unfastened to windscreen washer low. There is huge scope for making these more differentiated, personal and effective – for example offering a range of palettes, as most computer makers now do, or even letting people record and upload their own sounds, as with ring tones.

Many other products have sonic assets that they are well aware of. Cereal makers have used product sound for years. More than 70 years after their birth, Snap!™, Crackle!™ and Pop!™ are still selling Rice Krispies, while Martin Lindstrom reports that in his massive *BRANDsense* survey[95], carried out by Millward Brown, 74 per cent of those surveyed associate the word 'crunch' with Kellogg's – no accident, since Kellogg's worked with a Danish commercial music laboratory to engineer the particular crunch of their cornflakes. This kind of initiative may be moving one step further: in 2002 plant genetic modification specialists Monsanto announced they would be experimenting with increased levels of wax in the corn plant to create a GM crunch that survives saturation in milk.

There are also many products whose sound is a great asset that has not being fully utilised, or claimed by any specific brand – and in some cases it's being completely ignored.

One example of the former is bacon. This is not a product that excites much visually. Obviously the primary sensory component of the product experience is taste, closely followed by smell – but these can't be broadcast. The sound of sizzling bacon has been used in many TV ads, but no one brand of bacon has taken the Kellogg's approach and staked a claim to own this great asset. There is a tremendous opportunity here.

Other examples of more or less unexploited positive sound include the frisson of popping corks; the fresh and enticing crunch of many varieties of food (crisps, celery, apples, toast to name but a few); the satisfying snick of an airtight plastic container closing; the pop of bubble wrap; the click of a camera shutter (now recorded and played back when the button is pressed on a digital camera); the bubbling and hissing of a boiling kettle; the thwack of a golf club on a golf ball (or in cricket the famous sound of leather on willow, or in tennis the ping of a highly-strung racket on a new ball, or in various games of football the satisfying thud of a boot hitting a ball). No one brand has laid claim to any of these as far as I know, and yet they are all anchors for feelings of great satisfaction in the users of the product. Opportunities abound in the unexplored land of sound!

Product sound as a liability

Many products make sound that we consider unpleasant. In some cases the negative effect of this sound is felt only by the product user, but in many others it spills over and affects others – as with aircraft noise or leaf blowers.

At the extreme, making a nasty sound is what the product is for: smoke detectors, fire, burglar and car alarms, emergency vehicle sirens and car horns are all about making an intrusive and startling noise, using our deep wiring that links sudden sound to fight/flight hormone release. The sounds of these devices have been established over decades of research and practical testing all over the world. When I grew up in the 1960s, British police cars had just moved from bells to a two-tone siren (always a minor third – I wonder why?). I remember on my first visit to New York listening wide-eyed to the incessant sound of American sirens in the street far below – a sound I had only known from *Kojak* and *Ironside* until then. Now the two-tone siren has gone the way of the red squirrel in the UK, overrun and superseded by a hardier foreign import: the two-phase US-style siren, alternating wailing and warbling, has become a standard in most parts of the world, including Britain, despite the fact that some studies show the two-tone siren to be the most effective.[96] There is work afoot to improve the situation: in the UK the Noise Abatement Society is working to develop a broadband sound siren with greater directivity and less adverse impact on the health.

Most warning sounds are not so carefully designed. A huge number of products incorporate alarms, from microwaves to reversing lorries, and the vast majority of them are plain, boring beeps. Whenever I see a passenger cart at an airport or a railway station, with its constant beeping to warn people of its approach, I wonder how the driver stays sane. It's bad enough to hear that for a couple of minutes, but to live with it all day must be extraordinarily stressful. There are many sounds that would work as well but be much less unpleasant – for example a voice saying "excuse me", a melodic and soft-toned horn noise, or even a recording of someone clearing their throat triggered by a proximity sensor; one important practice for the poor operators would be to have a number of sounds so that there is some variety. This is a plea to all manufacturers: please try to make your warning sounds part of the product experience – and part of your brand. Be creative. Use your resources – agencies, brainstorming, designers – and see if you can't come up with something that warns people without beeping!

We've already mentioned the negative sound associated with cars (road noise in the cabin). Lexus has made a point of using its ultra-quiet interior as a major selling point to prove the quality of its luxury brands. There is a point of zero net benefit here, though. Lexus found that some customers were complaining they couldn't hear the sound of their own engine – a vital feedback signal when driving. Lexus is not alone in selling quietness: other car manufacturers, such as Ford and Chevrolet have joined a lengthening list that started many years ago with Rolls Royce.

Many other products whose inherent sound is a liability are well aware of the issue and are doing everything they can to optimise the user experience, and to minimise the fallout on bystanders.

Aeroplane manufacturers, encouraged by ever-stiffer legislation, have been making planes progressively quieter year by year. In the 1960s when a VC-10 passed overhead, conversation was not an option. Today's jets are radically less intrusive than that. The same is true inside the aircraft. Airbus coined the slogan "The quietest cabin in the sky" to market its A330 series, and the phrase is being used by airlines like Emirates when they sell their flying experience to potential customers. Some planes are even named for their sonic properties: the Q in the name of Bombardier's Q400 turboprop planes stands for quiet.

Computer manufacturers have been offering ultra-quiet PCs for years, though this has felt a little like the early days of organic food: the devices are often tucked away, hard to find and premium-priced. The rise of large, fast solid-state memory should see noisy disk drives disappear from many machines; also, some peripherals such as printers and keyboards are selling on the basis of how quiet they are, all of which should start to improve the auditory environment in offices. The large manufacturers' efforts are not enough for many people, and there is a profusion of websites offering tips on the quietest machines, and on ways to customise machines to make them even quieter; www.quietpc. com and www.silentpcreview are good examples.

In the home, we are surrounded by devices that make sound as a by-product of their primary function. Dishwashers, washing machines, tumble dryers, vacuum cleaners, hair dryers, hot water pumps, power showers, fridges, freezers, food processors, waste disposal units, fan heaters – our homes are full of sound, and most of it is an undesigned side effect of a thing doing what it needs to do; a necessary evil.

This is starting to change. Sound is becoming a major factor in product choice for most of these categories – but are the manufacturers aware of just how major? Danish design and testing specialist Delta reports on its website[97] that it tested several hair dryers. A panel rated the products separately on sight and on their recorded sound. The product that came top when mute turned out to have the worst-rated sound (described as cheap, annoying and unpleasant) and moved from top to bottom of the ratings when sound was included in the assessment. Bad sound can destroy the finest design work in the other senses.

Some device manufacturers have responded to this changing sensibility in their customer base. Washing machines, dishwashers and tumble driers commonly display decibel scores alongside their environmental ratings. Effectiveness and price are no longer the only considerations when people buy such things: they must now be as quiet as possible and minimise their environmental impact. I expect noise to become a key selling point for all the other devices listed above as well. In some cases this may take time and re-education: one manufacturer of leaf blowers found that reducing the noise of the product made customers think it had become less powerful.

Far too many manufacturers still seem blissfully unaware, or are unconcerned, about the auditory pollution they create. My personal *bête noire* is the food or drink chiller cabinet.*● In sound audits for various clients in food shops and in café environments we have measured the noise coming from these at over 70 dB. The freon compressor pump in these monstrous noise pollution generators is creating a hum that resonates in a big hollow metal box with plenty of parts that are free to rattle and vibrate in sympathy. This is only allowed to happen because no customer has asked: "Can we have a quiet one please?" The effects can be most unpleasant. One example was a visit to the head office of the Royal Bank of Scotland in London. In their hugely impressive entrance, a great hall with echoing acoustics created by stone everywhere, the dominant sound was a nasty, persistent buzz. I took my host (the then Director of Brand Strategy) to investigate, and we found a café on a mezzanine floor below the main hall with a single chiller cabinet producing enough sound to fill the whole space. If this had been a bad smell it would have been dealt with immediately. I have had this experience over and over with open-plan cafés in large office buildings, from the corporate headquarters of Vodafone in Newbury to the otherwise beautiful City Hall building that houses the Mayor of London and the Greater London Authority. Many commercial catering outlets – including pretty much every coffee bar I have ever experienced – have exactly the same problem with chiller cabinet noise pollution. If your company owns and operates chiller cabinets, please upgrade your specifications so that next time you order them you ensure they generate no more than 40 dB at one metre. The world will be a much better place for it.

A second class of prevalent noisemaker is coffee machines. I am not advocating returning to the bad old days of powdered coffee: the espresso machine is a great enhancement to the quality of life for millions of people. But why does it have to be in the same space as the customers? It generates over 80 dB of noise from roaring steam while it works, not to mention the aggressive banging of the filter that most

---

* ● You can enjoy some classic chiller cabinets on the website, along with samples of coffee machines and heavy diesels.

baristas feel is an indispensable element of the process. The racket makes the phrase 'a quiet coffee' a distant memory. It is not unavoidable. The surface used for banging filters could be heavily rubberised to remove at least that sound; also the machine could be located out back (many restaurants already do this) or surrounded by sound-absorbing panels to stop reflection.

A third, and more serious, class of products with largely unconscious negative sound is heavy vehicles. The combination of large diesel engines, resonant hollow metal and plenty of rattling attachments make these some of the loudest single products we encounter on a regular basis. At one metre, a large diesel vehicle can generate more than 100 dB of noise. Some of this is unavoidable until we develop a viable alternative to the diesel engine, but much of it is plain thoughtless – as with chiller cabinets, the fact is that quietness has never been high in the design specification. There are noise emission laws to be sure, and the limits set in them have reduced over the last 30 years, but the level of policing is minimal and the increasing levels of congestion (with resulting stop-start noise) mean that urban noise continues to rise. The UK's Noise Abatement Society launched a major campaign in 2007 called 'Silent Approach™' (in association with the Freight Transport Association), which provides guidelines and standards allowing for near-silent night deliveries (for example to supermarkets in residential areas). This eases congestion at peak times without creating any noise disturbance. It's achieved by the hauliers investing in new, quieter tools, such as plastic and rubber-wheeled trolleys and quiet vehicles – and of course in training their staff to be whisper-quiet!

In London the *Sounder City* plan, written by Max Dixon, showed the way to a quieter future, with tighter noise controls, investment in quieter vehicle technologies and resulting initiatives like the London Hydrogen Partnership, which operates clean and quiet buses in the city. I hope this plan survives political change and becomes a model for all major cities in the future. I also urge any heavy vehicle operators to reduce their noise emission specifications. As well as being seen to care about the quality and politeness of their drivers, there is excellent reputation and public image mileage to be had in organisations being seen to care about the noise pollution they emit. Maybe we'll start to see "How's my noise?" stickers next to the "How's my driving?" ones.

For years, marketers have been purveying products and services that saved us precious time – convenience foods, computers, more efficient domestic devices. Now, people are realising that quality of time is more important than quantity: no matter how many TV dinners we eat, we are still in a rush. I expect to see many more products and services, particularly at the luxury end of the market, selling qualities of time, such as peace, tranquillity, serenity or quietness, instead of quantity of time saved. Perhaps peace is the new time.

3.9 Soundscapes

Existing and potential customers are what commercial organisations exist for. Most organisations set high standards of customer service, and design spaces that customers will use in order to create the best possible experience – for their eyes. Unfortunately, customers' ears have been almost completely ignored, and the result is that the overwhelming majority of customer spaces have inappropriate soundscapes and are less effective than they should be as a consequence. Let's take a tour.

Corporate receptions

First impressions are powerful and can be hard to change. Most of us form an opinion about a person, brand, company or place almost instantaneously. Online research in Canada[98] found that it takes only 50 milliseconds for people to decide if they like a website or not. That is a purely visual experience, and it may take us slightly longer to form an impression in the physical world where all five major senses are in play – but we all know how fast we make our minds up, assimilating countless bits of data and forming an opinion without any conscious thought process to speak of. First opinions matter so much because, having formed them, most people then start gathering evidence to support them, dismissing or ignoring anything that doesn't fit.

When someone visits an organisation, whether they have come to sell or to buy, they transform from a relative neutral into an advocate or a critic. As someone who actually has direct, physical experience of the organisation they will pass their opinion on to others, who in turn will pass the opinion on to still more people, and so on. Every direct contact is a pebble dropped in the pool of reputation.

Receptions are there to receive visitors, by definition. They should also welcome people, offer useful communication and initiate a strong and positive relationship. It's sad that most of them fail to do any of this effectively. So many receptions just try to be impressive – to awe people rather than engage with them. This is a throwback to Victorian corporate relationship building, where the aim was to create a satisfying feeling of kudos for customers and a healthy level of intimidation for suppliers. Many of today's architects seem to be preserving a Gothic approach to receptions (large and overbearing) and resisting the modern, Renaissance style of business (scaled to and centred around the individual).

I have yet to encounter a reception that uses sound well. The main background elements in most corporate reception soundscapes are inappropriate acoustics (at worst those of a cathedral) and air handling equipment. In terms of foreground sound, the clear winner in the UK is one of the world's most effective cuckoo brands: BSkyB. There is a received wisdom out there that says it's good to be up with the times, and that the best way of demonstrating this is to have one or more plasma screens in reception showing Sky News. But the result is that if you close your eyes in hundreds of lavishly appointed receptions the only brands you experience are Mr Murdoch's, and those of his advertisers – who may well be competitors of the branded space they are invading.

Another all-too popular foreground sound is inappropriate music, often chosen by the receptionist.

There are so many opportunities for positive sound in receptions. Pleasing ambience would be a good start, maybe using generative natural sound or appropriate musical elements (in accordance with BrandSound™ guidelines of course) on top of decent acoustics and quietened air handling kit.

Most organisations could show some of their wares with sound. It's strange that advertising agencies, who produce a lot of sound and vision for clients, often decide to bombard visitors with MTV on multiple plasmas, presumably to stress how hip they are. What a missed opportunity to show how good they are at their job! Almost any major consumer brand could run fascinating legacy audio and/or audio-visual material to show its history and development. Any kind of organisation could use sound to communicate useful information to visitors: most first-time callers would appreciate a three-minute introduction to the company, with interactive options to find out more about its management, ethos, current performance, strategies, marketing and so on. Remembering the first golden rule, we can use hypersonic speakers to target this kind of content only to those who want it in sound pools, or offer people headphones, with an interactive interface similar to a gallery or museum guide. Finally, sonic art is a rich and fast-growing field and large corporate receptions could start to sponsor interesting and appropriate sonic art installations to inspire, amuse or entertain people while they wait – while creating associative overtones of culture, sensitivity, creativity and iconoclasm. For more on sound art, see the section on public spaces.

Lifts and lobbies

We don't spend much time in lifts, but employees do experience them multiple times each day. The growing fashion to install news services with sound in lifts should be resisted: these are places where people have their last chance to prepare for important meetings, to reflect on actions arising, and to talk with colleagues. Distractions in this space are unnecessary and unproductive.

Please do pay attention to the quality and tone of automated lift announcements. They do have an effect: in the Chelsea and Westminster Hospital the automated female voice, though delivered through a poor quality system, is very positive, bursting with pleasure in announcing that "this lift is going up!". Concerned people visiting sick relatives probably gain a little something from that on the way up. More mundane automations can become insanely irritating when heard 20 times a day.

Lift lobbies are usually silent but need not be. In London's BOX we created a wonderful soundscape that bathes visitors as they step from the lift: a looped edit of a piece called *Sleep* by avant-garde composer Mileece – music which simultaneously soothes, fascinates and entrances, without dominating. Looped sound in a lift lobby doesn't need to be long, as average occupancy of this space is just a few seconds, but it can do a disproportionately large job in immediately communicating the right values – in the case of BOX, that this is no ordinary office space. Why are lobbies and corridors silent everywhere when there is so much potential for setting mood and tone with sound?

Toilets

In Japan, leading toilet producer Toto has sold over 500,000 of a product called Sound Princess (*Oto-Hime* in Japanese). Installed in ladies' toilets as standard equipment in many modern constructions, it produces the sound of running water at the touch of a button in each cubicle, in order to mask the sounds of nature created by the occupant.[99] This may seem comical in Western cultures, where people are brought up to laugh at lavatory noises, but in Japan it addresses a real cultural issue – women are genuinely embarrassed to be overheard – and a resulting environmental and economic problem – they had taken to flushing multiple times, wasting a great deal of water.

The most common sound in most toilets is that of noisy extractor fans or air handling equipment, along with hand dryers that seem to output more noise than air. Replacing this unpleasant soundscape with intentional and appropriate masking sound is probably a polite and helpful move anywhere in the world, regardless of toilet etiquette. The masking sound could be musical (subject to BrandSound™ guidelines) but the most obviously appropriate sound is running water, which most people find soothing and which is entirely in context in a bathroom.

Case study: BP

Tools: soundscape

BP's customer research established that satisfaction with its service stations was low – mainly due to the toilets. The solution was to come up with a radical redesign of the whole toilet experience called Five Star Bathrooms, using nature-based visuals and sound, and kept spotlessly clean.

Ogilvy coordinated the project and engaged The Sound Agency to create the sound. To complement the visuals of sunflowers, fields and tress, we installed a soundscape of forest ambience and birdsong.

People now come out of these toilets smiling, and customer satisfaction has risen by an incredible 50%.

You can hear a sample of the BP soundscape on the website. ●

Meeting rooms

Very few meeting rooms are fit for purpose when properly listened to. The prime sound requirements for any meeting space are a high signal to noise ratio (in particular good insulation from people and any other noise sources nearby), confidentiality (again, insulation) and good room acoustics so that multiple voices do not become overwhelming. All too often one or more of these features are absent, and the result can significantly affect the outcome of the meeting, as those present strain to listen over intrusive background noise, and nerves fray due to unpleasant reverberation.

At one extreme is the open-plan meeting space, or its next-door neighbour the curtained-off meeting area. I have endured meetings in sizeable companies in open or semi-open spaces located next to buzzy sales teams and it's hard to imagine anyone having a productive time without immense effort. Trying to negotiate or communicate complicated information while half-listening to someone's loud phone call to a favourite client about their incredible weekend is very wearing. Whatever organisations save in interior design costs by creating these desultory gestures towards proper meeting spaces, they will lose many times over in unproductive meetings, poor outcomes, stressed staff, bad decisions, gossip, conflict and lost clients.

In the middle we have the low-cost partitioned meeting room. Usually the thin partition extends only to a suspended polystyrene tile ceiling with a void above. Unfortunately for this set-up, sound is very good at going around, or in this case over, obstructions. There's no such thing as 90 per cent soundproof: recording studio designers will tell you that a keyhole invalidates all the soundproofing material in the world. Sound always finds a way through. With partitioned spaces like these, it just jumps over the top of the partition, which might as well not be there for all the privacy it affords. Partitions themselves are usually made of cheap plasterboard with no acoustic absorption properties, or of single-glazed glass, which reflects some frequencies while allowing others to pass through and absorbs almost no sound at all. As a result, meetings in these spaces are a little easier than open-plan but it only requires a loud presenter or a healthy debate next door to wreck a meeting and produce most of the same outcomes as above.

If meetings are important to your organisation, provide for them properly. Create meeting spaces with good insulation, which means acoustically effective walls that reach the real ceiling. Ideally employ a professional acoustician to give specifications to your architect or interior designer because every space is different. As a rule of thumb we would recommend that any meeting room is no noisier that NC 25, with NC 35 as an absolute maximum. The soundproofing performance of the room's walls, floor and ceiling required to achieve this standard will vary depending on the amount of noise trying to get in. As a guide, but I stress this should always be reviewed case by case, a meeting room located in a noisy office will probably need all its border surfaces to have a Sound Transmission Class performance rating of around 50 – that is, they attenuate any sound passing through them by 50 dB. This means that even a noisy outside office at, say, 75 dB becomes a comfortable ambient noise level of 25 dB in the meeting room.

Soundproofing is only half the battle: we still have to consider sound quality, which acousticians think of in terms of speech intelligibility. Apart from noise, the main enemy of intelligibility is reverberation. Inside a meeting room, the target should be a warm, dry acoustic; I suggest aiming for RT of around 0.3 seconds. To achieve this, you will need to avoid the widespread designer obsession with wood, stone and glass, all of which reflect the vast majority of the sound that hits them, creating confusing and nerve-jangling reverberations that will raise stress levels in every meeting. Start by having a carpet: all meeting rooms should be carpeted, whatever the rest of the office looks like. Next, consider the ceiling. All acousticians look up when they enter any space, because absorbing unwanted sound is all about square footage: the more sound absorbers they can install the better the result, and in most cases that means using the ceiling. Install acoustic tiles, or if that is impossible get an acoustician to make some acoustic panels and suspend them from the hard ceiling. If you still have too much reverberation, hang some more panels on the walls. They can be made to look like unpainted canvasses in any colour, and can be very attractive as well as soaking up unwanted sound. In extremis you can even treat the underside of the meeting table. Any sensible investment in creating acoustically appropriate meeting rooms will be repaid many times over through hundreds of much more effective meetings over the years.

Case study: The Box

Tools: office soundscape

EDS and the London School of Economics came together to create a truly innovative meeting space. Visionary author and business guru Louis Pinault wanted the Box to be a complete multi-sensory experience. He brought in The Sound Agency to create sound that would inspire, energise and support learning.

The Sound Agency created a whole range of innovative soundscapes. In the lift lobby, generative music from sonic artist Mileece soothes and intrigues. Inside the Box the default soundscape is subtle nature sound, which elevates energy levels and cognition while relaxing people physically. A 'Cabinet of Curiosities' is full of fascinating artefacts, many of which are sonified via pressure pads, infrared beams or other triggers, with the sound delivered in pools via hypersonic loudspeakers. The Sound Agency even installed music played by plants, using the electrical potential across their leaves to drive MIDI-based generative music. The installation required over a kilometre of Cat5 cable! The system is driven by a network of computers controlled by a custom-designed user interface on a Wifi tablet.

There was sound to remove, too: the noisy air conditioning system had to be silenced with bespoke masking boxes. Every detail of the space was sonified: even the door-open alarm is a cuckoo calling - which only the staff will notice.

Visitors to the Box report being able to work far longer and far more effectively without feelings of fatigue. Staff love the space.

You can hear some of the Box sound on the website. ●

Shops and other retail spaces

Appropriate, well-designed soundscapes can increase sales in shops by up to 50 per cent in some cases, and between 5 and 30 per cent as a rule. I know this because we have done it and measured the results in our work with major retailers.

Given that well-designed sound can achieve such outstanding results, it is heart breaking to enter the vast majority of shops and find the same mindless soundscape.*● There is a myth in the retail trade that lively pop music is the right soundscape for any purpose. Partly fostered by those who are selling piped music and partly by Western culture's discomfort with silence (probably pining for WWB, as I've suggested) and its need for constant stimulation, the practice is unfortunately becoming prevalent. In opposition there are pressure groups lobbying against any use of piped music, often incorrectly termed 'muzak'.[100]

Muzak is in fact a registered trademark of the Muzak Corporation, the world's longest-standing provider of piped music, starting with the playing of records over Tannoy systems to factory workers in the 1930s. Muzak estimates that over 100 million people now hear its music every day. It has some regrettable legacy, for example its use of tempo to 'program' people subliminally, repeatedly building from slow to a fast climax to stimulate productivity, and its aesthetic crimes of the 1970s and 1980s, the age of *Yesterday* on pan pipes in hotel lobbies and lifts. But it has moved a long way in recent years under new management, and is now starting to consider more carefully the appropriateness of the music it sells to its clients. It talks about 'audio architecture', and commissions research into music effectiveness. However, in 2009 the company went into Chapter 11, finally emerging a year later with a big funding package from a new majority shareholder and a new structure that divides content, interactive experience and hardware. Perhaps this heralds a more sophisticated approach to brand experience.

More maturity in this market is to be encouraged. Music is not homogeneous, nor is it the only sound that might be appropriate. Deciding to play music in a store without considering why, how and what kind, is like ordering the decorators to use paint without specifying the

---

*● On the website we have created a short mix of some of the unfortunate soundscapes we've encountered in high street shops.

type or colour. Every store has a huge range of soundscapes to choose from, just as it has a huge range of interior designs to choose from. It's tragic that so few spend any time at all thinking about which is the best soundscape for their specific room, brand, customers and location.

The reason this is important is that, as we saw in the SoundFlow™ section, there is strong, copious and reasonably unanimous evidence that music affects people's behaviour at the point of purchase. Pace of shopping, amount spent and product selection have all been shown to be affected by music. Sound can make or break brand relationships at point of sale. If sound is in phase with visuals and with the brand's character, it can create a huge boost for the brand's impact, but if it's out of phase it can cancel out visual aspects and create a net zero effect.

Where a retailer is selling its own branded goods, there is no excuse for getting this soundscape wrong – and yet time after time we have audited single-brand shops that have inappropriate, counter-productive and incongruous sound.

Where branded goods are sold by a third-party retailer, as is most often the case, the situation is more complex as the different brands may want conflicting soundscapes. This requires sensible negotiation by the brands with the retailer, which of course may itself have its own well-defined BrandSound™ that overrides any of these considerations. Even if that is true, directional speakers make it possible to create sound pools for point of sale displays that don't pollute the rest of the store with their noise.

Almost all the research that's been done in shops concerns the sound of music; little is known about the effects of other sounds (such as noise) on shoppers. From the research we do have, one message comes through loud and clear: music affects people powerfully when they are shopping, so it is vital to design each retail space's sound with care. A powerful effect can be negative as well as positive. People *notice* music, and as psychologist Helen Gavin says:

Retailers need to address the fact that people are experiencing negative effects of their use of music in their outlets. It is clear that the intention, in using music, is to satisfy the consumer in some way, but in fact that it can detract from any pleasurable experience. The aim of retailers and service providers, in the use of music, should be to reflect the atmosphere of stores, products, and services, generating a comfortable experience.[101]

I suspect that the current indiscriminate playing of music in almost every retail space is generally doing more harm than good. For every store that has chanced upon the ideal musical soundscape, there are many playing inappropriate, ineffective music that's causing many customers to walk out, not to buy more. As Professor Adrian North, the UK's leading expert on music and consumer behaviour, succinctly puts it: "No music would be better than the wrong music."[102] So, while Miss Selfridge (a brand aimed squarely at teenage girls) could spend money with a sound branding consultant only to arrive at the inevitable recommendation that the bubble-gum pop music they already play is the ideal soundscape for their customer base, most shops can and should review their soundscape with an expert. Most have not done this, which is why the sound in most shops is arbitrary, hostile and incongruous.

The four Golden Rules are paramount in getting retail sound right: all sound must be optional (or at least accurately targeted), appropriate, valuable and tested.

When we work with a retailer, the first task is usually to identify and remove inappropriate sound. As well as the current music, this may include hums and buzzes from machinery, street noise, reverberation due to poor acoustics and service noises like shelf-stackers dropping trays or the rattle of service trolleys. Playing music on top of this mess is like putting icing on mud: the result is never going to be a cake. *●

I have mentioned that we have measured chiller cabinets in supermarkets generating over 70 dB in buzz. Those particular cabinets looked great, with mist drifting down across the plump fruit and vegetables – but stand in an aisle between two of these monsters and all you want to do is leave the store. As we have seen, the most basic behavioural effect of sound is to attract or repel, and in this case the store in question is telling its customers to go as soon as possible. It's a classic case of schizophonia: the eyes are being told "come in, you're welcome, stay" but the ears are receiving "not comfortable, not safe, get away". Chiller cabinets are a regular offender everywhere from cafeterias and supermarkets to coffee bars. There is no reason why they need to be so noisy, and I fondly hope that this book may help bring about a new generation of whisper-quiet

---

* ● As well as the high street retail mix you will find on the website an even more impressive collage of typical and painful soundscapes all recorded in major supermarkets.

chiller cabinets. As mentioned in a previous section, a chiller should generate no more than 40 dB at one metre. Retail store managers should adjust their purchase order specifications accordingly.

Heating, ventilation and air-conditioning (HVAC) equipment is another major irritant in shops. Stand under a massive ceiling-suspended pump in one of the vast Tesco Extras and you can clearly feel the bowel-quivering bass frequencies. When we assessed the effectiveness of DIY chain Homebase's in-store radio station Homebase FM, we found that it was inaudible in many stores due to the noisy HVAC plant. As with chiller cabinets, customers may not be conscious of this noise, but the effect will nevertheless be to encourage them to leave the shop as soon as possible. Again, the solution is to re-specify, demanding kit on the next refit that throws out no more than 40 dB of noise.

Another consistent offender in many shops, and again particularly in supermarkets, is the humble trolley. In supermarkets, as in major transport termini, broken trolleys are often the single biggest source of noise. The traditional wire-framed design with rattling plastic wheels is a guaranteed noise generator, even before someone knocks a bearing or axle out of true to create yet another lame trolley with that familiar ker-chink, ker-chink sound. There is usually nobody to take these out of service until they stop moving, and so they shout out their crippled misery to everyone within earshot for months, even years.

We need a new generation of trolleys, with plastic bodies and rubber wheels on silent bearings. Only when retailers understand the damage that's being done, even though customers are unconscious of the effect, will this come to pass. All too often I have come across retailers who take the view that nobody has complained, so nothing needs to be done. I believe that the high level of competition in retail will leave these ostriches head in sand at the rear, while more enlightened operators steal their customers by actively marketing peaceful shopping. In Europe, it's already common to see relatively silent plastic and rubber trolleys in hypermarkets. I hope competition will see this fashion quickly spread around the world.

Large service trolleys are even worse: I have measured these generating 80 dB of noise as they go past.*● They are designed to carry goods as

---

* ● One such, encountered in a high-quality mall, is captured for posterity on the website.

cheaply and efficiently as possible, but their utilitarian makers have failed to think about their effect on people's ears. Even behind the scenes, their crashing and clattering creates stress for those using them. But when they make their appearances in front of house, the customers are yet again being told subliminally to go somewhere else, somewhere safe. If you are using trolleys like these, I can only suggest that you order quiet ones next time, if not now. The standard as always should be 40 dB at one metre.

Penultimate in this little rogues' gallery is street noise. The modern high street is uncomfortably noisy, at 65-80 dB. Many shops have a policy of leaving their front doors open all day to remove any barriers that might deter customers from stepping in to browse. This may work in a quiet street, but when the shop doors open onto a busy main road, the result is to admit frightening levels of aural (not to mention chemical) pollution. I have stood in booth-like cell-phone outlets on Oxford Street wondering aghast how the people who work in there survive without therapy, and how any sensible business conversations take place at all. To add insult to injury, some of them are playing semi-audible music, presumably in the hope that this will make everything fine. It doesn't.

Even in large shops on major roads, the front 5-10 metres is a pretty uncomfortable place to be, particularly if there is a wooden floor that reflects most of the street noise instead of absorbing it. It would be interesting to look at the sales per square metre for such shops. Though there are, of course, many factors at play here, including product visibility, footfall rates, and so on, I would predict that sales are often much higher in the more comfortable areas at the back of the store than in these noisy zones by the doors.

Doors, of course, are barriers to leaving as well as to entering. I hope that shops on busy streets will at least test closing their front doors, or installing automatic closing, to see what effect it has on sales. I suspect that the old-fashioned shop front with revolving doors was a more effective solution than the current open-door fashion, but of course revolving doors are not compatible with modern access requirements, so the best option is simply to shut the door – which has major benefits quite apart from making the shop more pleasant for the ears. In the UK a campaign called Close The Door was started by a group of women in Cambridge to lobby retailers with hot air blowers (in the winter) and air conditioning (in the summer) pouring expensive energy out through wide open doors into the street outside. The campaign spread like wildfire and is now working with Government departments; several major retailers have adopted the closed-door policy and report no adverse effect on sales at all. We can but hope this will spread globally.

Finally in this parade of problems we have acoustics. As well as catering for fashion victims, many shops are themselves the victims of interior fashion, with designs that look great and sound awful. Unfortunately, architects and interior designers spend only a tiny fraction of their training learning about sound, and in my experience almost none of them see it as the fascinating and flexible extra material it can be for them. Acoustics become important to architects and interior designers only when the space being designed is all about sound, such as a concert hall or recording studio, in which case specialist firms are brought in. As far as I am aware, nobody is yet specializing in the acoustics of shops, which is a real shame for us customers.

Most bad acoustics in shops result from a combination of practicality and aesthetics. Hard surfaces are easy to clean, so they are used in most catering and high-traffic places. At the same time, most of the materials that designers and architects favour for their looks are hard: prime among them are wood, glass and metal. Carpet wears out, is harder to keep clean and is currently unfashionable. Until it comes back in, we may have to live with hard floors.

As usual, an acoustician will look up to solve the problem: if the floor and walls have to be live (bright and reflective), at least the ceiling can work hard to absorb sound. Either absorption panels or acoustic tiles in a suspended ceiling will improve the sound radically when specified by an expert. Better still, if designers can be persuaded to use some soft materials the result will be less chaos and more calm.

Once undesirable sound has been minimised, and only then, new sound can be introduced. This is where we consult BrandSound™ guidelines: if you don't have any, pause here until you have employed a specialist consultant to create some for you. From the BGs will come clear direction for the right background and foreground elements of your soundscape. These will differ from shop to shop, determined as always by the brand, the customer demographics and psychographics, the location, the acoustics of the individual space and so on.

Background elements of the soundscape might be natural – birdsong, running water and other popular natural sounds are very soothing for people escaping from traffic-dominated busy high streets – or musical, comprising either a streaming of carefully-selected recorded tracks that match the BrandSound™ guidelines or generative musical compositions. Silence is always an option: there is no vacuum to fill here, and in many cases a decorous hush is more conducive to shopper wellbeing than any sound could be.

Foreground elements of the soundscape might include announcements for customers or staff, promotional messages and warning sounds. Please invest enough in the public address/voice address system to achieve good quality in these announcements, and have it calibrated by a professional so that all sources of announcements are delivered at the same volume. If your staff are going to make announcements that customers can hear, train them to do it well and enhance your brand in the process.

For consistency over the whole sound system, there are clever pieces of technology called autogain units that can automatically alter the volume level to maintain a constant amount of headroom over the ambient noise in the store. I recommend using them rather than leaving the volume control to any member of staff. I have come across the latter in several retail situations and it is far too prone to abuse. If one person's favourite track comes on they may turn the system right up and then leave it there, inappropriately deafening customers; alternatively if the store gets very busy the carefully designed soundscape may be drowned by ambient noise and all the staff may be too busy to notice. A far better approach is to calibrate the system with a target signal-to-noise ratio in dB (a typical amount of headroom would be about 5 dB) and then set an autogain unit to deliver this all the time. If need be, give the manager alone the authority to override the default in special circumstances, for example a late night stock check where the staff need to be motivated and entertained with a bit of a party atmosphere.

In-store promotional messaging is a whole different can of worms. There is a fast-growing trend towards video screens in retail stores. This flourishing industry makes an apparently irresistible offer to retailers: install the screens and advertisers will pay top dollar to get sales messages to shoppers at the point of sale. In theory, the system becomes a profit-centre for the retailer as well as boosting sales for the featured products (by up to 30 per cent, depending on the research you use.) Promotional material for in-store TV systems usually end with a throwaway claim that customers are happier too, supported by one or two quotes from delighted punters. This anecdotal approach makes me suspicious, and UK research firm Shoppercentric confirms there is cause to be so. Its research shows that shoppers do not understand the role of these TV screens, and two thirds of them pay no attention to them. Shoppercentric's Managing Director, Danielle Pinnington, says:

> "Shopper needs are rarely being considered in the design, development and use of in-store networks... communication in-store should primarily be aimed at driving action, and will only meet its full potential if it is in tune with shoppers and aligned with the shopping process."

It's outside the scope of this book to indulge in a full discussion of public TV (also known as digital out of home (DOOH), in-store media and a variety of other names). What does concern us is a Samsung Screen Survey[103] that revealed that 47 per cent of these screens have audio switched on. If retailers install screens with sound as yet another unplanned element in an already chaotic soundscape, it will have a negative effect on dwell time and thus sales, and this will at least partially, if not fully, offset any gains the products featured on the TV screens might experience. The net effect of the two influences remains to be calculated by researchers.

This new intrusion emphasises how the whole of a soundscape needs to be considered, and constantly reviewed.

One important force we've not yet considered is the staff. Customers come and go in a matter of a few minutes, but staff cannot flee the scene. The effect of an unpleasant soundscape on them accumulates over time and is not to be underestimated. Our discussion of the three Cs in the section on noise concluded that control – or lack of it – is a key factor in sound creating stress, unsociability and even illness. Shop staff increasingly have no control over the music piped into their workplace, and the effect can be serious. Let's not forget that MOR music has been used as a weapon by US military and security forces (against the Viet Cong and the Waco cult respectively).

Any piece of music has a tipping point beyond which repetition becomes painful. For complex music that point takes some time to reach, but for simple three-minute pop songs it is perilously close, regardless of personal taste. As with eating too much chocolate, excessive repetition can put people off even their favourite songs; when the repeated track is one they are at best tolerating to start with, the pain can start almost immediately.

According to a UK survey in 2005 by Retailchoice.com, 31 per cent of shop workers endured the same album between six and 20 times a week – and 16 per cent had to bear more than 20 repetitions a week. Britney Spears was voted most irritating artist, followed by Usher, Kylie Minogue and 50 Cent.

Like so much else, the whole thing gets intensified at Christmas, when the UK's Royal National Institute for the Deaf estimates that some stores play Jingle Bells over 300 times. BBC News magazine reported one shopworker's sad tale, which should serve as a torchbearer for all the rest:

> Iona, 33, recalls her time at the Chester branch of a toiletries chain, which had Pavarotti's Christmas CD on a loop from mid-November to January.
>
> "I went mad, and was constantly accosted by lovely old grannies asking about the music, saying wasn't it wonderful to work in such an environment. I was reduced to dribbling like a child and smiling politely.
>
> "This nightmare of music was compounded on Christmas Day – my one day off – when it turned out someone had bought my mother the CD. It was the first thing that went on the player. I cannot listen to Pavarotti to this day."

In Austria in December 2003, shop workers threatened to strike, claiming that the 'psychological terror' of repetitive Christmas music was driving them to become aggressive and confrontational.[104] The mild mutiny spread to Germany and Holland before dissolving.

With so much music available in the world, it seems criminal to subject retail staff (and shoppers) to death by repetition. If your BrandSound™ guidelines lead you to music as a background element in the soundscape for your store, please make sure that the playlist is large enough to avoid any noticeable repetition – as a rule of thumb, at least three times longer in duration than the time period you play it over each day, with automatic shuffling so that the songs don't occur in the same order every time. The piping and streaming companies can all help with this, installing satellite-linked or Internet-linked hard disk players that can store and randomise thousands of tracks.

Once you have identified the right soundscape in theory, it's vital to test it. We have carried out such tests for major retailers, which is how we know that appropriate, well-designed soundscapes can increase retail sales by up to 50 per cent – although this level of increase is available only where the starting point is a very poor-sounding space. We are confident of achieving increases of 5 to 10 per cent in the average retail outlet with properly designed sound.

There have been claims that jolly pop music increases retail sales, but these usually turn out on closer inspection to be small samples of customers saying that they like the music. This is not what I mean by effective testing: our methodology for it is given in the section on the Golden Rules of sound in Part 2. We focus on sales as a measure because if the soundscape is more pleasant, dwell time (average length of stay in the store) will increase, and so will sales – simply because people are more comfortable. All things being equal, increased sales due to a new soundscape therefore indicate increased dwell time which itself implies increased customer satisfaction.

Once testing has validated (or maybe improved) your retail soundscape, it should be redone at least annually to keep abreast of market and customer changes. There may not be such a thing as a perfect installed soundscape, but we can nevertheless patiently and continuously strive for improvement.

Showrooms

Showrooms and shops are very similar, so most of the same principles apply. Showrooms' acoustics can be poor due to the amount of glass and marble used, and there is often a plasma TV with news or sport polluting the soundscape. When a brand spends millions to look its very best, it's particularly sad that it allows other brands to pollute and even dominate its soundscape.

There is huge scope for creating showroom soundscapes that enrich the brand experience rather than diluting it. Carefully chosen licensed music would be a start, but so much more is possible. Specially commissioned music (written to a good brief derived from BrandSound™ guidelines) can support and amplify a brand in its signature spaces. Sound designers and sonic artists can also contribute here, if carefully commissioned to create a unique custom soundscape that will make the showroom into an unforgettable experience. Impressive technologies such as hypersonic loudspeakers, flat surface transducers, 3D sound and generative systems can all help to deliver a rich, powerful and positive experience.

Catering and hospitality

Hotels

Once notorious for dreadful piped music in lifts, hotels have come a long way in the last twenty years. However, there are still some which play old-style substandard cover versions of pop songs, as exemplified by the recording on the website accompanying this book, made in the Marriott Berkeley Square and featuring probably the worst version of *Hey Jude* ever recorded.*●

The rise of the boutique hotel has changed the sound of hotels for good, and music is now a major element of the brand experience for many hotels. Paris's Hôtel Costes has shown what can be accomplished here. In the 1990s its cool bar benefited nightly from ever-cooler music played by house DJ Stéphane Pompougnac, and the sound became a destination. People came from all over to enjoy Pompougnac's mixes of super-cool lounge and chill-out music. Due to the number of requests, the hotel released a trial compilation CD in 1999. Eleven albums and millions of copies later, the series has defined a new territory for location-based compilations.

Most boutique hotels can't match these heights, but they do take care to choose their music and its style to reflect the flavour and individuality of their brand. In recent years the mainstream hotel chains, always alive to the new tricks of the boutiques, have adopted music styling to help create their own identities.

My plea to the hotel trade, though, is to apply the essential principles of psychoacoustics – to run SoundFlow™ in planning mode, defining

---

* ● I'm delighted to say that we captured some of this rich cultural experience and have included it on the website for you to enjoy.

the effects that are desirable in every space, be it a lobby, bar, restaurant, ballroom, lift, spa or guest room, working out what sounds can deliver those effects and then employing professionals to design soundscapes that deliver exactly what is required. Sound is too powerful to leave it to any individual's taste to determine what gets played, even if that person is the General Manager.

When they arrive in a hotel room, most people rearrange the furniture or otherwise make their own mark, claiming the territory by making it personal. Until a few years ago, the only sound that could be used in this personalisation was the radio or TV. More recently, boutiques and now many mainstream hotels have started putting CD players into rooms. Now these are being replaced with sockets for guests' iPods. The inevitable next stage is music streaming on demand. I believe that another popular option will be soundscape streaming, so that every time you visit a hotel you will be able to switch on your favourite soundscape (whether natural, musical, generative, or some combination). As life becomes more interactive, hotels can start accepting contributed soundscapes from their guests, and even paying royalties in the form of free accommodation if they become popular and are used by many other guests.

One final concept for hotels is to create stunning sound installations in high profile public spaces. We installed birdsong in the lobby of the London InterContinental Hotel because it is reassuring, comforting and familiar for people entering an unknown space for the first time when tired after long travel. That soundscape is delivered in mono, but how impressive it would be to encounter a piece of superb sonic art, maybe generative and based on the natural sounds of the hotel's location in full 3D sound in a lobby or atrium space. People love sound design like this; here is a major opportunity for hotels, which (like all other businesses) have for so long focused just on how they look and missed out on the potential of sound to enhance their brand and their customer experience.

Restaurants

We've seen from the research quoted in the SoundFlow™ section that music can have a strong effect on people's behaviour in restaurants: fast tempos cause them to eat and drink more quickly and to spend less time and money in the restaurant; slow tempos encourage longer meals and higher expenditure results, especially in drinks. We also know from research that classical music can cause people to choose more expensive wines by creating an ambience of quality. Finally, we know that texture has an effect, particularly style: studies have shown that people spend more in restaurants when listening to classical or pop music than when listening to easy listening or no music.[105] Presumably in all these conditions, people stay longer and spend more because they enjoy the ambience more, so everyone wins.

And yet we must pause here before suggesting that upmarket restaurants should all play slow classical music to maximise sales. Few soundscapes arouse such strong feelings as those in restaurants. Many restaurant critics are vituperative adversaries of the whole idea of music while you eat, and will mark a restaurant down if it dares to permit the practice. Whilst critics' views may not necessarily reflect the views of the general public, they will often reflect the views of their readers, who may include current or desired clients. It's important for a restaurant to assess its clientele carefully before moving from the relatively safe haven of discreet silence.

The filters in SoundFlow™ are particularly important in an environment which is about far more than just getting fed: most people in a restaurant are there in pursuit of pleasure, and every aspect of the experience needs to be designed to work with all the others to create a consistent, effective whole that gives the restaurant its individual character. The environment – acoustics, sound sources and sound system – and the nature of the restaurant brand (or the values of the person or organisation behind it) are critical.

Every case will be different, so I shall simply warn against both dogma and imitation. Music can make a dining experience just as it can break one. Some restaurants (Hard Rock Café, Buddha Bar) would be incomplete without their music. Others would be diminished by background music. In between is a huge grey area where far too many restaurants are what marketers call me-too: they play music simply because other restaurants like them do so. It is in this grey area where important questions need to be asked: How can music express our brand? How can it help differentiate us? What can music say about us? What do our clientele really prefer? Does music add value to their experience? Please, where restaurant music is concerned, don't 'just do it'. Use the tools in this book to make a personal decision that suits you and your customers.

Further down the quality scale, mindless music is at least matched as a problem by grim acoustics. Cafés, fast food restaurants and many mid-range establishments are designed with cost and convenience in mind, which means hard surfaces and lots of reverberation. This can really spoil the dining experience. I urge architects and interior designers – and their clients – to involve expert help from the start and to design the sound just as carefully as they design every other aspect of the space. * ●

Bars

A friend of mine walked into an empty bar some time ago for a short business meeting with a colleague. As soon as the bar staff saw them, they turned the music up to the point where orders had to be shouted and conversation was extremely difficult. My friend had to insist before they would turn the music down, and when they finally did and communication became manageable he asked them why they had done what they did. They replied that it was the policy of the management that the music be at that level whenever there were any customers in the place. Perhaps the management have seen the research quoted in the SoundFlow™ section that linked loud music to increased drink purchasing. If so, they need to take care to consider this result in its proper context. It is certainly not an absolute truth. Each bar should be able to respond to the needs of its customers in accordance with the Golden Rules.

The kind of authoritarian jollification encountered by my friend is all too common. As usual it rests on the assumption that music is essential, and the secondary assumption that for extra atmosphere the music should be loud. Buzz and volume are widely, and wrongly, treated as the same thing. I have been in bars with a great atmosphere at the level of an intense murmur, where music would have been a rude intruder. At the other extreme, the liveliest bars, for example those around sports venues on match days, are wasting their time playing music: it just interferes with everyone shouting at each other.

The regrettable rising tide of unconscious public music (and TV) is very evident in pubs and bars. Jukeboxes at least gave the customers control over the selections, and were often confined to one room out

---

* ● Please have another listen to the noisy café samples on the website to be reminded of just how painful this kind of soundscape can be.

of several. Now most pubs and bars play radio, or have a jukebox, or subscribe to a piped music service, or show sport on a big screen, and the sound is audible everywhere. As usual, my issue with this is that nobody has properly planned the soundscape. While it may be appropriate for pub or bar chains to make their brand experience consistent with music, the invasion of quiet village pubs and character bars by piped music and TV sport is often tragic.

The critical factor for bars to consider is the role of conversation. In a meeting place for young people, awkwardness is eased when conversation is limited by loud music; in a local watering hole, people want to talk and listen. This dynamic changes throughout the day, and most pubs and bars need to create soundscapes that change as the time passes to reflect this.

I remember a wonderful pub near Cambridge called the Tickell Arms where the jodhpur-wearing aristocratic landlord played Wagner at high volume on Sunday lunchtimes to go with the roast and the strawberries. It was perfect. In some cases, a bar's music will define its whole identity. Just as Hôtel Costes made its global reputation through the music played there, so did another famous Paris establishment, the Buddha Bar. DJ Claude Challe created a popular mix of global and lounge sounds and, two years after this bar/restaurant opened in 1997, it, too, launched a hugely successful compilation CD series that now includes 12 primary and several secondary albums. Challe has moved on and Ravin taken over, but the vibe remains the same and the continuing global momentum generated for the brand by its musical manifestations has led to Buddha Bars being opened in Kiev, São Paulo, New York, Dubai and Beirut among others.

The art of doing this well is to decide on who the customers are and then to employ a creative professional (usually a DJ) to programme music that will attract them, using the principles of psychoacoustics as a background to inform (but not dictate) the creative selection. Too many bars play music for the sake of it, with no thought as to its selection or its effect. At its best, bar music can be the heart of the brand: at its worst, it can destroy a place people loved. As always, we must tread with care.

Staff spaces

Noise at work

Some occupations require people to work in noisy environments: construction, heavy industry and heavy engineering are obvious examples. The standard global limit for acceptable noise in any workplace is 85 dB. Above this level, prolonged exposure will inevitably damage hearing, so employers are obliged by law to offer protection and regular hearing tests and employees are legally required to use the protection provided. Tragically, few do. The onset of hearing damage is gradual and invisible, like ice being eroded from beneath – but once we fall through, there is no way back. Noise-induced hearing loss is irreparable. Much more must be done to educate people because they simply cannot perceive the damage that's being done to them – so they do not see the need to take precautions until it's far too late.

The modern office environment

The past few decades have brought many new fashions to the office workplace: informality, outsourcing, home working, hot-desking, team working – and, most of all, open-plan environments, which have all but eliminated closed office settings, particularly in knowledge working industries where the watchwords have become flexibility and communication.

The danger is that the baby will be thrown out with the bathwater. Closed offices certainly isolated people or teams, but they were also quiet and largely free from external distraction. The open-plan office is much noisier. But does this matter? Are the benefits so great that the drawbacks are insignificant?

A study in the US by the Yankelovich Partners, carried out for the American Society of Interior Designers (ASID), found that 70 per cent of office workers believed their productivity would be higher if their environment was less noisy. And yet a follow-up study for ASID found that managers were largely unaware of noise as a problem, with 81 per cent of them unconcerned about office noise. Might this be because they sit in their own quiet offices?

Several trends are creating noisier offices:

- Open-plan layouts
- Higher workstation densities
- Team areas which encourage group and interpersonal interaction in open office, often collocated with personal working space for flexibility
- Reduced height furniture systems to encourage communication and establish eye lines (which are also ear lines)
- Interior design that leaves ceilings, walls and even floors uncovered and thus reflective
- Larger computer screens, which reflect more sound
- Increased use of speaker-phones, which encourage loud speech.

It comes as no surprise, then, that 71 per cent of office workers rate noise as the number one distraction in their workplace. But is their subjective perception in line with objective research or is it just a case of people always having something to complain about? *●

Noise and productivity: the research

A series of studies both in the laboratory and in the workplace have established beyond any doubt that office noise is now the primary cause of productivity loss in offices. A while ago I commissioned Professor Adrian North, one of the UK's leading authorities on the psychology of sound, to review all the scientific findings on the impact of office noise. His four general conclusions were:

1. The great majority of the research shows that office noise has negative effects on employees.
2. Office noise is related to reduced job satisfaction and dislike of the office environment.
3. Office noise increases stress among employees.
4. The negative effect of office noise is exacerbated when employees do not believe that they can control it, when it is perceived as unnecessary, or when employees have not been exposed to office noise previously.

---

* ● There's some typical office sound on the website. As always, when listening without seeing, the nature of the soundscape is much more clearly revealed.

Because this problem is so little recognised, I want to emphasise it by lifting as direct quotes some of the most significant conclusions of the papers Adrian North reviewed in that study.

"Distraction was most strongly related to degree of self-control of the noise and noise predictability. The most critical noise sources for the annoyance response were other machines than those used by oneself, whereas telephone signals had the largest effect on distraction."[106]

"Office noise and lack of privacy affected worker satisfaction and mental health."[107]

"Noisy office-like conditions affected long-term memory recall and self-reports of mood. Being able to hear irrelevant conversations impeded ability to learn."[108]

"Office noise increased workers' levels of epinephrine (the neurotransmitter indicative of stress); reduced the number of attempts made to solve puzzles (which was interpreted as evidence of lower motivation); and led to fewer attempts to make ergonomic postural adjustments to computer work stations. It did not lead to greater perceived stress."[109]

"Although the office did create a favorable social climate, this did not offset S[ubject]s' negative reactions to work conditions but rather appeared to exacerbate the problems. Consequently, no evidence was found to support the claim for improved productivity in open-plan."[110]

"Irrelevant speech may contribute to mental workload and result in poor performance, stress, and fatigue."[111]

"Office noise led to worse performance on a measure of 'integrative complexity' and a 'simple cognitive task' and led to perceptions of greater disturbance and stress. These effects could be ameliorated by masking white noise, but performance was best when there was no noise."[112]

"Loud, uncontrollable office noise led to worsened mood (and a more negative mindset) and increased tension. These effects were not found when either the noise was controllable or when it was quiet."[113]

"Office workers preferred quiet music or nature sounds to a completely quiet office and these helped to mask background noise."[114]

Not surprisingly, office noise is correlated with increased sick leave and staff turnover, and with reduced willingness to help co-workers. Most startling of all the results is the one we noted in the section on the cognitive effect of sound, where Banbury & Berry (1998)[115] found that normal office noise reduces effectiveness in cognitive tests to just *one third* of its level in quiet surroundings. The tests were designed to simulate standard knowledge work – tasks that require the use of short-term memory and concept manipulation. The idea that many office workers are operating at one third efficiency due to the noise around them should give pause for thought to even the most ardent advocates of the modern open-plan flexible workspace.

These studies show that there is a major problem here. They underline the importance of the three C's (control, contrast and conversation) and they call into question the whole received wisdom that open-plan offices are more efficient. It's interesting that you can fit roughly three times as many people into an open-plan office as into a traditional office with many rooms – and if they are knowledge workers, this research shows that they become one third as productive. I propose a number called Treasure's Constant, which is the amount of work you can get out of any given building. You can achieve it with more people working less productively (open-plan) or with less people working more productively (traditional); what you can't do is increase it by pushing density up.

Variations in work and in individuals

There is a spectrum of types of work in relation to the degree to which noise will diminish productivity. At one end of the scale is monotonous, repetitive and intellectually undemanding work, where we often find people playing music in order to alleviate their boredom. Here, extraneous sound can actually aid performance, which is why Muzak Corporation et al make a good living out of piping music into places of work. Similarly, creative teams, where the work is about being in flow, or sales teams, where buzz is vital, may do better with music or the sound of co-workers at a high level.

At the other end of the scale lies what is generally termed knowledge work: office work that is largely intellectual, non-collaborative and that involves structured thinking, planning, problem-solving, designing, attention to detail, processing large numbers of words, numbers or other symbols. In this case, sound is highly destructive to performance.

We must also remember that people vary widely in their susceptibility to noise interference. For many, highly sensitive to sound, generally easily distracted, or simply used to quiet, noise is a major issue. Others may find it relatively easy to carry on working despite the sound around them. As usual, the real test is what people do, not what they say. I know teenagers who claim they can do their homework better with loud music playing. All the evidence I know of says this is a form of denial practiced in the interest of having more fun. There's nothing wrong with doing homework with music on – but it will take longer and the lessons will not be as easily learned as they would in quiet surroundings where attention could be completely focused on the task at hand. Similarly in offices, as so often with sound, many people will not be conscious of the negative effect on them of surrounding noise, because they are so used to suppressing it from their consciousness.

The research gives us general rules, but it is important to consider these individual and situational variations before jumping to sweeping conclusions (for example that all noise is bad). We learned in our discussion of noise that one person's noise is another person's signal. I have often encountered serious conflict in advertising agencies that have thoughtlessly placed creative teams next to 'suits'. Creative people seek flow, contact with their muse, and music is often just the catharsis they need. As a result they prefer to work with music on all the time. In contrast, the knowledge workers in the account handling team, struggling to complete a major client presentation on time, cannot function with noise going on. Put the two side by side and you guarantee friction. It is vital when planning an office to try and group people by their type of work so that their ideal soundscape is similar and this does not happen.

Privacy and communication

To assess whether you have a problem in your own offices, you can do three things. First, you can survey your staff. Notwithstanding what I have said about so much of sound's effect being unconscious, the US research we have quoted shows that people in offices are often all too aware of the noise around them and how it is stopping them from working effectively. It's hard to measure effects until you start making major changes; asking people what they experience is a simple and low-cost start. Second, you can measure the level of privacy they are experiencing. There are three recognised levels of office privacy:

Confidential privacy: co-workers can overhear muffled words but the meaning of the spoken message is not intelligible and they are not distracted from their own work

Normal privacy: some sentences are intelligible to co-workers but the volume level of the speech is not distracting to them and they can generally continue to work on their tasks.

Transitional privacy: co-workers can overhear most words, most sentences are intelligible and distracting to them; their concentration is disrupted, stress results and work performance is significantly decreased.

Normal privacy is the minimum standard for any knowledge working office.

Privacy is inversely related to speech intelligibility, for which as we saw in the section on acoustics, the industry standard measure is speech transmission index, or STI. The table below, derived from work done by the American Society of Interior Designers, shows the inverse relationship. Transitional privacy is a combination of the bottom two categories of privacy (marginal/poor and no privacy): if your office's STI is above 0.2 then you have a problem.

Privacy level	Privacy Index	STI	Subjective intelligibility
Confidential	1.00	0.00	Bad
	0.95	0.05	
Normal, non-intrusive	0.90	0.10	
	0.85	0.15	
	0.80	0.20	
Marginal/poor	0.75	0.25	
	0.70	0.30	
	0.65	0.35	Poor
	0.60	0.40	
	0.55	0.45	
	0.50	0.50	Fair
	0.45	0.55	
	0.40	0.60	
	0.35	0.65	Good
	0.30	0.70	
No privacy	0.25	0.75	Excellent
	0.20	0.80	
	0.15	0.85	
	0.10	0.90	
	0.05	0.95	
	0.00	1.00	

STI can be reduced by masking sound, which we discussed in the section on acoustics. It is always best to employ an acoustician and let them use the ABC of acoustics as appropriate to improve matters. Best of all, start by employing office designers who understand sound, and have an acoustician on the team from the very start of designing a new office.

I recognise, of course, that some people like to work in open plan offices for social reasons, or in order to bounce ideas off colleagues and solicit ad hoc advice and assistance. However, they too will need to be able to concentrate and focus as a regular part of their job, and the same principles continue to apply. Managers and office designer must plan and control the precise degree of open-plan interactivity they wish to create, and must optimise the availability of quiet space within the open plan environment.

Modern offices typically have work space and meeting space (as well as common areas). If we must have open plan, I believe planners need to start creating a third category: quiet space. The number of people camping out in meeting rooms or working from home because they have a report to write should be a clue to the need for such spaces which, like public libraries, offer peace and quiet for those who need it.

Health clubs / gyms

It's not just fashion that drives the playing of loud, up-tempo music in health clubs and gyms, or causes so many people to wear personal stereos while jogging. Researchers have found that, when listening to music, people can exercise at sub-maximal rates for longer periods.[116] The mechanisms are believed to include dissociation (engaging with the music so much that we simply don't notice the messages from our muscles as much as we do when that's the only major input we have); straightforward arousal; and also synchronisation of music with exercise. We don't need scientists to tell us that it's also just more enjoyable to exercise to rhythmic music.

However, as usual with sound, intelligence, care and sensitivity are required, along with a little vital knowledge. That knowledge starts with the fact that prolonged exposure to 85 dB or more of noise causes permanent hearing damage – and many gyms and in particular aerobics-type classes pump out over 100 dB. Sustained for an hour, this is damaging the hearing of everyone present, and as for the teachers the damage is likely to show in just a few years as serious noise-induced hearing loss (NIHL).

Guidelines from the American Council for Exercise suggest that classes should be held with a sound pressure level (SPL) no higher than 70-80 dB. If you are in the sports business, please operate within these levels so that you are not robbing people of one of their most precious assets. If you are a gym member and you believe the music is louder than a typical street, either ask your gym to turn the music down or take earplugs. If you run with a personal stereo, please check the section of this book on personal soundscapes before setting the volume level.

Spas

Like most people, I love spas. Big fluffy robes, herbal teas, the smell of essential oils, the hands of an expert – but what is that terrible music?

I am probably more sensitive to sound than most people, and more critical of soundscapes too, but even allowing for this I have visited far too many spas on my travels where the music has been barely tolerable, even downright irritating.

Some spas have inadequate systems, like a small tape recorder (still!) or boom box in the corner playing the same cassette or CD over and over. Even where systems are built in, they often have low-grade ceiling speakers that grudgingly offer up only the barest band of harsh mid-range frequencies.

And then there's the content. Unfortunately there is apparently an unlimited quantity of new age music on acoustic guitar, piano or flute coming out of California, but it might as well all be one track because it all sounds the same to me. Some experiences have been laughably incongruous: not long ago I was subjected to bagpipes and drums – sounds that were designed for war, not peace and relaxation!

For anyone who runs spas, I suggest making some important changes to the traditional model. First, offer choices. Different people have different tastes and you should offer at least the following three choices: carefully chosen classical music, contemporary music and nature sound. Second, have the music programmed by an expert, not by the people working in the spa. They are experts in massage, not music. Third, ensure there is enough music that it never repeats during a single session – and preferably not during the same day.

There is plenty of good music out there. In Dubai I recently came across Fridrik Karlsson's *Feel Good* collection, which is high quality and specifically designed for this application. Even better would be a generative installation of gentle, floating musical and natural sounds: a soft aural waterfall, always flowing and fresh but at the same time always essentially the same. In a soundscape like that, people could *really* relax.

Public spaces

In the section on noise, we discussed in some depth the problem of increasing ambient noise levels, particularly in cities, and the vast cost it creates for society. We need not repeat that discussion here. We can, however, say on an optimistic note that people are starting to pay attention to the issue of social sound. In many countries sonic art is becoming increasingly prevalent in public spaces.[117] In this section we'll touch briefly on the potential of sonic art from a business perspective, then review some public spaces that have an effect on the commercial world: schools and universities, hospitals and public transport.

Sonic art

There is a centuries-old tradition of public sonic art in Japan, where many public gardens used to feature a *suikinkutsu*, an elaborate water feature that exists purely to create the sound of drops of water resonating inside a buried pot. The thought of such a feature in a modern urban public garden is ludicrous: the subtle, delicate sound coming from under the ground would never be heard over the traffic noise, even by those few not wearing their iPods while jogging.

We may have lost some of the subtler qualities of sonic art but happily its quantity is on the increase. There has been a boom in sound installations over the last decade, and interest seems to be continuing at a high level. I hope this heralds a new appreciation for the concept of soundscape design in public spaces.

Sonic art is generally very popular. In my experience most people find sound fascinating, so playing with it in the cause of art creates great interest, as recent installations in London's Tate Modern and New York's Whitney Biennial have proved. Sonic installations are still much rarer than other forms of art in our public spaces, so they also have novelty value.

They can be hugely imaginative and engaging, partly because of this novelty effect, and partly because sound as public art is still so relatively unexplored. The Greyworld group, founded by Andrew Shoben in Paris in the 1990s, has been a leading force in exploring sound-based interactions in public spaces around the world. They started by tuning a set of Parisian railings so that, when a stick was run along them, they play *The Girl From Ipanema*. They moved on to install a blue carpet on Dublin's Millennium Bridge, with inbuilt pressure sensors triggering the sound of crunching snow or swooshing leaves as people walked. A work called Trace was installed in the famous maze at Hampton Court Palace;

people wandering in the maze are engaged by faint voices and sound effects apparently left behind by other visitors from previous centuries. My favourite is *Bins and Benches*, installed in 2006 outside the Junction Theatre in Cambridge, UK: in this public square the bins and benches move (benches huddle under tress when it rains, and bins line up on Wednesdays waiting to be emptied) – and delightfully they sing joyfully when the sun comes out, barbershop from the bins and a soprano choir from the benches. The whole thing is solar-powered. Greyworld shows us just how brilliantly sound can enhance any public space, bringing surprise and delight into our lives when we least expect it, and inviting us all to play with our environment instead of merely enduring it.

For business, sonic art presents at least two opportunities. First, there is the opportunity to sponsor sound art installations in prominent public spaces, creating an association with a vibrant and impactful art form and a sense of original thinking. Second, there is the opportunity to commission and install sonic art in customer or staff spaces, in corporate receptions, atria or other suitable areas. With the new technologies available to the sonic art community (see section 2.4 above) these installations can surprise and delight without interfering with the functions of a space.

Education

Classrooms in any educational establishment should have a reverberation time (RT) of less than 0.5 seconds, but many have at least double this. Revised building regulations and standards such as the UK's BB93 are addressing the RT problem for new builds, but older schools still form the bulk of the stock. Many schools are also located under flight paths or next to railway tracks or major roads, although research has shown that noise from these sources directly impairs performance.[118]

These acoustic defects both work to reduce speech intelligibility (SI) so that large numbers of students in our schools and universities hear at best one word in four of what's being taught. This is dramatically repressing educational standards, as well as teaching young people how *not* to focus.

Both these effects are bad for business, which depends on bright and well-educated young people for its future, but what can be done? Fortunately, there are actions that business can take. Lobbying governments, via business or trade associations or personal connections, is one course of action. Another is to sponsor programmes of acoustic improvement in local schools or universities, especially where these already feed into your workforce. A third is to get involved with specialist hearing organisations such as the US League for the Hard of Hearing or the UK Royal National Institute for the Deaf. They have well-developed lobbying and research programmes, and as they represent partially-deaf pupils as well as the profoundly deaf, they are particularly concerned with the effects of poor SI, which are felt even more profoundly by those with impaired hearing.

Hospitals

There is a whole book to write on this one topic. I had the misfortune to spend a fair amount of time in hospitals some time ago while visiting seriously ill family members, and whilst my admiration for the people who work there is limitless, I despair at the soundscapes they – and more importantly the sick patients – are expected to endure. Intensive care units in particular offer nothing but hissing and beeping – the sounds of lives on the edge, of existence hanging by the thread of technology. According to a 2006 study by Johns Hopkins University, hospital sound levels have increased from an average of 57 dB in 1960 to over 70 dB today. People get well in these places despite the soundscape, never because of it.

We know that sound can play a major role in healing. Music is being widely used to treat or ameliorate dementia, mental disorders, chronic pain, stress, depression, insomnia and a variety of specific complaints from asthma to ulcers.[119] Conversely, it's clear that the terrible and intimidating sound in many hospital areas is working squarely and powerfully against recovery. Falk and Woods (1973) measured noise in various typical locations including by patients' heads, in recovery rooms and in baby incubators and concluded:

> On the basis of present knowledge of the physiologic effects of noise, these noise levels probably stimulate the hypophyseal-adrenocortical axis of patients, exceed the noise threshold for peripheral vasoconstriction, pose a threat to hearing in patients receiving aminoglycosidic antibiotics and are incompatible with sleep.[120]

Quite apart from the personal, emotional and social consequences, this affects business : if better soundscapes were creating more effective hospitals, less time and productivity would be lost due to employee sickness.

Much more research needs to be done in this area. First, we need to quantify the problem and the potential benefits of solving it. Then we need to take action: find and install acoustic countermeasures which are medically acceptable and will not, for example, become bacteria traps; adopt new working practices (such as vibrating pagers and warning devices instead of the default beeps); and design and install healing sound, delivered to patients (even while unconscious) through lightweight disposable headphones. Or through ambient diffuse sound systems using surface transducers.

The best way business can help is to sponsor research into the area. The end result will justify the expense, but if further payback is required the PR benefits of being seen to care in an innovative and different way about people's health could be very substantial.

Public transport

Public transport is a major noise polluter. Aeroplanes, trains and the diesel engines of buses all factor highly in the annoyance league tables for people outside them; they also affect people inside them.

Planes have become much quieter in the last twenty years, but then there are far more of them now. Many major airports, start flights before 5am and do not stop until after 11pm. As long as flying continues to get cheaper and the demand continues to grow, there is no prospect for frequency being reduced and we can only support the efforts of the manufacturers to make even quieter jets.

For those inside the fuselage, soundproofing has improved continuously to the point where the best planes hardly require a raised voice. With noise-cancelling headphones, flying need create no problems at all for even those with the most sensitive hearing. This shows what can be achieve where there is a will and a sufficient budget.

On the roads, the diesel problem is only just starting to be tackled in forward-thinking cities like London, with its ambient noise strategy and trials of innovative quiet buses. The vast majority of the buses in the world are highly toxic in terms of noise (not mention carbon emissions). On board, the noise is moderate to bad, but for pedestrians and cyclists it is often atrocious, and is a major contributor to the urban cortisol and adrenaline overdosing problems we have discussed earlier in this book. We can only look forward to at-source noise-cancelling technology bearing rich fruit, and quickly. In the meantime we should also actively support trials of alternative power systems for commercial vehicles, be they hydrogen or electric or hybrids. Urban authorities must start to factor noise into their plans for road-based public transport in the future.

Trains are equally poor. The noise outside them is severe, and new generations of commuter trains are, in my experience, even noisier than older stock, often creating terrifying squeals from their wheels and penetrating whines from their great electric motors. Inside, they are much better soundproofed than they used to be, so the experience of travelling on modern trains is acceptable or even pleasant, but on platforms or in locations adjoining the track the noise seems worse and more frequent than before. There is unfortunately little that can be done to ameliorate the sound of metal on metal as the wheels grip and rattle on the track – unless and until magnetic levitation trains break out from their current niches and become commonplace.

Going underground, all these problems are compounded by reverberation. There is nowhere for the sound to go, so although there is no fallout on nearby homeowners or pedestrians, the effect on travellers can be intense. The Paris Metro is exemplary, with its rubber-wheeled trains, large double tunnels and spacious stations. London's Underground is right at the other end of the scale, with antiquated rolling stock, metal wheels, snug fitting one-train tunnels and platforms that collect the sound, letting none escape. The SPL inside the cars on the Victoria line can exceed 90 dB, and the profile of the noise from wheel squeals, track and engines is appalling. I believe drivers and platform staff should be wearing hearing protection as they are often being exposed to noise well over the WHO recommended maximum of 85 dB – as are regular passengers. For London Underground customers, I strongly recommend travelling with hearing protection until the necessary billions are spent on renovating this creaking system. And please note: hearing protection does not mean a personal stereo turned up loud enough to drown out the noise: that is just accelerating the hearing damage. (Please see the section on personal soundscapes for more detail on how to protect your hearing against noise.)

Away from the vehicles themselves, we have the sound of massive termini to contend with. The sheer size of these buildings makes them acoustically difficult. Many have RTs that would not disgrace a cathedral, and they are full of people and noise. Sadly, public address systems are not the highest priority when it comes to budget time, so many of these great buildings are surviving with PAs that should have been replaced decades ago. On top of lousy acoustics and cheap, poorly sited loudspeakers, many termini compound the problem by using staff who are untrained in public speaking, or who have strong accents. The combination of these factors all too often gives rise to indecipherable announcements that just add to the noise and as a result create greater customer frustration.

Help is at hand with new PA technology, such as the products we discussed at the end of the second part of this book. All that's required is the budget to replace the old systems and the will to select and train staff explicitly for their intelligibility and their fit with the brand voice; then we will have effective communication.

Underneath that, we can create soundscapes for these spaces that are designed to relax people. There is often stress associated with travelling. Slow-paced, gentle, soothing sound can create an ambience that defuses rising anxiety and calms everyone down (instead of the current effect, which is to wind people up). Clearly, given the acoustics, such soundscapes need to be slow-changing, low density, and based on familiar sounds that create reassurance. The soundscapes we created for BAA were like this and they were well received by passengers, who found them calming. I do too, and I often play them at home when I want to relax or work quietly.*●

---

* ● There are two extended sections of these ever-changing, generative soundscapes on the website (one for day, one for night) so that you can do them same. Thanks to BAA for allowing me to use them here.

Case study: BAA | Heathrow and Glasgow Airports

Tools: soundscape audit, soundscape

In 2005, BAA was seeking to improve customer experience in all its airports. It retained The Sound Agency to find out what was happening on the ground. We carried out an intensive audit of all four then existing terminals at Heathrow, covering every space open to customers in each of them. A comprehensive report contained many quick wins and showed that there was a major opportunity to improve brand consistency and brand experience by taking control of the soundscapes in these spaces – from the largest issues to small and yet significant ones like squeaking trolleys.

Following the Heathrow audit we designed a soundscape for trials in Glasgow Airport. Its aim was to improve passenger experience by reducing stress levels. We employed a daytime generative soundscape that included birdsong and ambient musical elements at a very slow tempo, and a nighttime version that had gentle water instead of birds (which do not sing at night) and solo instrumentation instead of pads to create a feeling of companionship and intimacy; both were played in the public walkways and lounges at very low level. Many passengers were unaware of the sound; those who did notice it were very positive in research. Whether people noticed the sound or not, it had a significant effect: the best indicator of its success in relaxing people and slowing them down is that sales in the airport shops went up by 3-10% when the sound was playing.

You can hear a sample of the Glasgow soundscape on the website. ●

Private spaces

There are three spaces we like to think we own and in which we have at least partial control over the soundscape: our home, our vehicle and, in a recent development, our personal soundscape. Business is affected by all three, and itself affects them in turn.

Home soundscapes

One day maybe it will be true to write that we consciously design and create our home soundscapes just as we do our colour schemes and furniture. That is far from the truth today.

In the first place, very few people are conscious of their home soundscape, or of the fact that they could design it. For most people, the ways they mould their home soundscape are unconscious, often mixed up with other factors – for example, switching on the TV or the radio for comfort and companionship, or listening to music. The other internal components of a home's soundscape are usually subliminal and left alone: creaking floorboards; groaning pipes; whirring, humming or buzzing pumps, fridges and other domestic appliances. We are generally unaware of these things but together they make up a 'soundprint' that we know well: we recognise foreign sounds instantly, even to the extent that they will wake us from deep sleep while louder constituents of our domestic soundprint will not disturb us.

As we saw in the section on product sound, there are signs that people are becoming more noise conscious, that peace and quiet are becoming highly desirable commodities in the home: it's common now for washing machines, fridges, vacuum cleaners and tumble dryers to publish their noise output in decibels alongside their environmental rating. However, the vast majority of people are unaware of – or just put up with – the sound that's created inside their home.

In the second place, our home is usually not our castle when it comes to sound. Most of us are invaded by outside noise, whether it be aircraft, traffic, railways, industry, construction, neighbours, dogs or services like refuse collection. The noise we make and impose on one another grows every year and the problem is worldwide. In the average home today, peace is a pipedream.

Reclaiming the home soundscape?

Even before the advent of recorded music there have always been ways to alter and improve our home's soundscape (for example fountains, crackling fires, wind chimes or caged birds). Some cultures are more sound-conscious than others: in Japan, many features of both house and garden have traditionally been designed to make or permit small, subtle sounds that combine to create a harmonious whole. Rice paper walls were created for teahouses in order to allow occupants to hear the whole building rather than just the room they were in.

Today's ways of adapting our domestic soundscape are less refined and deliberate, and somehow more subversive and threatening as a result. In the nineteenth century came the sounds of machinery, with the industrial revolution's steam and gas introducing a whole new realm of domestic noise, such as hissing gas lights, creaking hot water pipes and roaring boilers. In the twentieth century came recorded music, electricity and broadcasting, and the mainly human-made soundscape of the old days was gone for ever as our houses started to resonate to strangers' voices, music and sound effects on radio and gramophone, culminating in the spread and now almost universal dominance of TV.

Domestic silence is rare, especially in cities. When I visit my wife's hometown in Italy's Dolomite mountains, I am always enthralled by the quiet both inside and outside her mother's house; like most city dwellers who visit the country, I find real quiet a startling change. I know some people who find the peace of the countryside unbearable, so conditioned are they to the urban soundscape. Many people feel uncomfortable on entering their silent home, and their first action is to switch on some source of sound for company.

As with domestic appliances, there are signs that people are starting to become more conscious about sound, and to show an interest in alternatives to music or TV as a domestic soundscapes. In Germany a few years ago there was a fast-selling CD aimed at sole occupiers; it featured the sounds of someone else cooking, showering and generally keeping you company in your solitary home. More and more hotels offer alternatives to traditional alarm clocks, waking you with a wide variety of nature sounds instead of the news or music or some strident alarm. There is a big potential market in home soundscapes, and it may not be long before we all get home, switch on the lights and choose our soundscape for the evening.

Excitingly, companies like London-based Future Acoustic are developing interactive sound for home as well as commercial locations. Not only generative, but also responding to what goes on around it, interactive sound opens up a whole new world of creative, endlessly changing, appropriate sound for the home. We commissioned soundscapes for Future Acoustic's prototype system from sound artists Scanner and David Toop, and the effects are very encouraging. We also commissioned a programme from sound designer Paul Weir that is specifically designed to act as a modern-day lullaby; once it senses you are asleep, it changes function to mask intrusive noise during the night, helping you to stay asleep instead of being awoken.

In the future, carefully designed sound generated inside the home may help combat the effects of unwanted sound intruding from outside.

Vehicle soundscapes

Engineers make great efforts to control the sound intruding into our cars from air friction, from the engine and from tyre friction on the road. The goal seems to be complete silence, though when this is approached we can find it inappropriate: we do need feedback from the machine we're driving to control it properly.

We have already considered the product sound of the car itself. Over the top of that, the soundscape we create inside our private car tends to be a microcosm of the soundscape we create in our house, though sometimes more extreme. The cult of the big sound system, with subwoofers large enough to bounce the whole car up and down on its suspension, is almost entirely male and could be classed with many other flamboyant courting or aggressive territorial displays found in nature. One assumes the owners of these cars do not customise their homes in quite the same way – fortunately for their neighbours.

For the rest of us, the typical soundscape is radio talk or music, and/or conversation with occupants or on the phone (hopefully via a legal hands-free kit). It's interesting that the destructive effects of conversation on concentration and possibly on reaction times have not been well researched. If the results of research in workplaces turn out to apply in cars, any form of conversation, whether on the telephone, to another person in the car or on the radio, will be found to degrade our ability to think and perform complex cognition (such as route finding or careful, anticipative town driving) but music will improve our morale and possibly energy while performing boring or repetitive tasks (such as motorway driving).

Music also affects our driving through physiological entrainment. It's more difficult to be stressed when sitting in a traffic jam with Miles Davis's *Kind Of Blue* or Barber's *Adagio* playing; by the same token it's hard to drive at 30 miles per hour with Led Zeppelin's *Heartbreaker* or Underworld's *King Of Snake* pumping out of the stereo. The first two are low tempo, reflective pieces, calming and reassuring us, while the last two are high tempo and exciting, demanding to be played at high volume and calling forth adrenaline in large measure. Unless we are very well trained, our driving will reflect our physiological and emotional state: stress, excitement, fear, anger and the fight/flight hormones will tend to make us more aggressive, selfish and fast, while a calm, serene, secure and confident condition will tend to slow us down and make us more yielding and socially conscious.

Little research exists on this topic but early indications are that there is a link between aggressive music and driving anger.[121] More research is clearly needed. It may be that in years to come playing music or listening to a chat show while driving will be as frowned upon by society as talking on a mobile phone is today. It is likely that many accidents happen every day because someone was distracted by sound (for example violently disagreeing with someone on a chat show) or because their behaviour was altered by sound (driving too fast because of adrenalising music), and yet more because the driver was fumbling with controls on some sort of sound device. There are unlikely ever to be statistics showing the link between sound and driving: you can't test someone for sound abuse!

The car is one more private soundscape where consciously designed sound probably has much to offer. Sounds specifically designed for the car, delivered generatively so they never bore the user and reacting to road, traffic and driver condition as well as to the time of day, could be a powerful aid to safety, as well as being a lovely experience.

For business, this soundscape matters because all of the above is just as valid for drivers of most commercial vehicles as it is for us in our private cars. Badly driven commercial vehicles cost more to run, create bad public relations ("How's my driving?") and damaged reputations, and can cause catastrophic accidents, especially where public transport is concerned.

We've all had the experience of getting into a minicab with deafening music playing. It does not reflect well on the cab firm and the best will not allow it because it's insensitive and puts the customer last, not first. The same effect is created by a liveried van with loud music issuing from the windows, whether parked to deliver or driving aggressively: the brand of the company whose livery is on the vehicle is tainted by the lack of consideration shown by the driver to the public.

Every organisation needs a policy about music and radios in its commercial vehicles, especially where these come into contact with the public. They are moving brand experiences, and inappropriate sound has an effect just as much as belching black exhaust smoke, poor driving, inconsiderate parking or simply obvious lack of cleaning and maintenance.

Personal soundscapes

Personal soundscapes are a recent creation. Short of the children's practice of sticking our fingers in our ears and humming, none of us was able to create a mobile personal soundscape until the arrival of the TPS-L2 Sony Walkman in 1979. Now personal soundscapes are becoming the norm, particularly among the young. In 2003, 35 per cent of the UK population owned a personal stereo, with MP3 at that time an insignificant factor. Now digital has done to CD what CD did to tape: my children have never bought a CD, and think that all music comes from computers, tablets and especially mobile devices. According to mobile industry pundit Tomi Ahonen, there were 4.4 billion mobile phones in use in 2010, with sales that year alone of 1.37 billion; coverage in the industrialised countries was 88 per cent, while in the developing countries it was 59 per cent. 60 per cent of these devices play audio. The reach figure of 4.4 billion for mobiles compares to 4.2 billion for radio, the same for TV, and just 900 million for home PC-based Internet.[122] Publicis CEO Maurice Lévy clearly had an inside track when he said in 2006: "In a couple of years, most of the information you share, most of the advertising you read, most of the messages you send, most of the music you listen to will transit through your cell phone."[123]

The music industry is finally reinventing itself to supply this new way most people listen to music – digital, mobile and personal. Services like Spotify, last.fm, Pandora, Grooveshark and We7 make listening to anything, anywhere a breeze – not to mention the profusion of illegal sources and the rampant file-sharing that are testing the old model of copyright and creative ownership to destruction.

Aside from the vile, antisocial habit known as 'sodcasting' – the playing of loud, distorted music through the tiny loudspeakers in mobile handsets by youths on buses and trains – most of this easily-accessible music is being listened to through headphones. The result, much discussed in sociological and psychological circles, is that our public shared spaces are effectively fracturing into a profusion of small mobile personal spaces.

This epidemic of ear buds can be seen as a defensive response to the increasing clamour of modern living. Sussex University's Dr Michael Bull believes that one of the main drivers behind the popularity of personal stereos is people's desire to regain some active control over their own space, particularly in urban environments where we are assailed by constant visual messaging and high levels of ambient noise. The simple fact that people have chosen what they are listening to makes them feel less dominated – we are back to the three Cs again, with control alleviating stress. Other motivations include subverting unwanted visual messaging, transforming unpleasant or mundane activities, removing feelings of responsibility for events in the immediate vicinity, and (particularly for women) deflecting unwelcome attention.

As we've already noted, the personal stereo has brought with it a new relationship with music, which is now the soundtrack to many people's daily activities, and is spread more broadly and yet more thinly than ever before. Hundreds of millions people are listening to it for billions of person-hours a week, but most are experiencing it as a secondary input while they do something else, like read, walk or work.

This is not the only major effect of the personal soundscape. With the rise of the personal stereo have come grave concerns about a potentially disastrous increase in noise-induced hearing loss (NIHL). As we've seen when discussing noise, 85 dB is generally acknowledged to be the sound level beyond which one's exposure should be limited, with recommended exposure times in Europe starting at 8 hours for 85 dB and halving with every 3 dB increase. (Remember, each increase of 3 dB is a doubling in sound energy.) At 100 dB, the recommended maximum exposure time is just 15 minutes.

Although disputes continue about how to measure the output of personal stereos, most experts agree that the majority produce well over 100dB at high volume settings, the effect of which is usually compounded by being delivered deep into the ear by bud-type headphones. Although few people probably start at such high volumes, the ear/brain mechanism adjusts almost like a compressor, making any given volume seem less impactful after a time. This habituation means that people tend to edge the volume up at regular intervals, ending up at a much higher level than they would choose from a standing start. If you can hear the sibilant sounds of hi-hats from someone else's personal stereo, they are almost certainly damaging their hearing. The general rule of thumb for users is that you should be able to hear someone talking to you from a foot or two away while you listen to your music. If you can't, your hearing is in danger.

For young people today, danger is becoming damage. In 1998 Niskar et al published a paper showing that around 15 per cent of American teenagers have permanent hearing damage.[124] In his book *The Power of Sound*, Joshua Leeds reports that between the seventh and twelfth grades the average American teenager listens to 10,500 hours of rock music – just slightly less than the entire number of hours spent in the classroom from kindergarten through high school. Leeds also notes that 35 per cent of the 30 million Americans suffering from hearing loss are victims of NIHL, as opposed to the effects of ageing – and, most disturbingly, that a university study found that 61 per cent of college freshmen exhibit some hearing loss.

It's likely that much of this damage in young people results from personal stereo abuse. Other causes are loud gigs and clubs, though these have been around for many years and there's no evidence to suggest that their effects are growing. It is almost certain that the damage is now growing exponentially with the massive uptake of personal stereos and the lack of guidance about the effects of abusing them.

This wouldn't be such a concern if it weren't for the fact that NIHL is irreversible: hearing does not convalesce or regenerate after being injured. We can only start to imagine the future social and economic cost of this very recent phenomenon. Severe hearing loss makes communication very difficult, and often causes isolation and depression as a result. For any economy, the effect on productivity and economic activity of half an entire generation going partially deaf would be nothing short of catastrophic.

For those who are determined to take control of their own sound, there are alternatives to loud music delivered through bud headphones. High quality, full cup headphones can satisfy at lower volume levels because they have a wider frequency response and they keep out more of the ambient noise. The downside is that they are bulky and more expensive, so they don't fit well with the young person's budget or demand for instant convenience.

Music is not the only personal soundscape to consider. Active noise cancellation (ANC) technology has been around for some time and the technology is now so good that on trains and planes I often wear my Bose noise cancelling headphones with ANC turned on, but without music playing, to create a quieter experience.

In some extreme environments, the right personal soundscape option is effective hearing protection. As mentioned previously, European Community legislation stipulates that employers must provide hearing protection in ambient noise levels greater than 85 dB, and that employees must *wear* such protection in ambient noise levels greater than 90dB. But many of us experience much greater noise levels in our daily lives: trains, traffic, building sites, night clubs, gigs – such noise sources generating between 100 dB and 140 dB can all create permanent damage, particularly if you are already skating on thin ice after years of hearing abuse.

As a long-time drummer, I am keenly aware that if the ice breaks there is no going back, so I have invested in high quality ear protection for gigs. These custom-fitted earplugs are flat attenuating – in other words I hear everything perfectly but 25 dB more quietly. With attenuation ranging from a mild 9 dB to a hefty 35 dB, such plugs can just make urban sound far less threatening: for people working in severe noise they can save your hearing.[*]

---

[*]   These are soft silicone custom-fitted earplugs supplied by Advanced Communication Solutions, who make such ear protection for the likes of Grand Prix racing drivers (with inbuilt radio speakers) and rock stars (with inbuilt monitor speakers). ACS can be found at www. hearingprotection.co.uk.

Unless and until the world becomes quieter (both aurally *and* visually) people are going to want their own auditory space. Maybe soon we will carry around a range of pleasing soundscapes to shield us from the noise of modern living, delivered through high quality lightweight ANC headphones at moderate volume.

Meanwhile, and for the foreseeable future, governments and voluntary organisations need to make every effort to persuade personal stereo users to watch their volume – as in the UK Royal National Institute for the Deaf's 2005 'Don't lose the music' campaign. Business should support these initiatives, because if they don't work we will encounter a drastically reduced supply of healthy labour. The baby boom is over in the Western economies. Unlike the Arab world with its massive bulge of young people, the Western countries have an ageing population and a narrowing pipeline of employable people to deal with in the next 20 years. We can scarcely afford for somewhere between 20 and 50 per cent of them to arrive at work with damaged hearing, compromised in their ability to communicate or to use the telephone. This is what will happen unless action is taken to educate young people about NIHL and the dangers of ear bud headphones in particular.

The web

Sound is coming to the Internet, and it will be a major competitive differentiator for those who use it well.

The web will find its voice: the parallels with the development of sound in other dynamic media are clear. Cinemas started silent, then developed low-bandwidth analogue sound on small speakers; then there was the exciting discovery of stereo sound through big speakers; then Dolby 5.1 and full-scale sound design. This journey from silent to low-fi to hi-fi to surround applies equally to home music systems, TV, radio, PCs, in-car entertainment, handhelds – you can plot where each technology is today on the curve.

Today's Internet is still largely silent, with a minority of low-fi sites and a handful of hi-fi pathfinders. The reasons for this – low bandwidth, clunky players and tiny loudspeakers – are largely historical. Now broadband has achieved critical mass in most of the developed countries. Flash and other embedded players were a start, but the coming of HTML5 at last allows audio to be written into the code of a static web page, which is a major breakthrough. Many home computers nestle among multi-speaker set-ups, so most websites browsed at home are already missing a trick by ignoring the power of sound. And for business users, whether in the office or on the move, Bluetooth and stereo headsets already allow them to enjoy web sound without bothering those around them.

What nobody predicted was the much more rapid rise (especially in the developing countries) of the mobile web. Computers are fast being outpaced by mobile phones, whether smart or not, as web browsing devices, and since most of those phones are already being used to play music with headphones attached, mobile web sound is a natural and yet largely unexplored medium.

Businesses that sell sound (or products that make a sound) can reap immediate benefits from web sound, so these tend to be the sites that lead the way: Amazon invented risk-free music buying by offering track samples; Ferrari lets you revel in the sounds of its cars while you dream of owning one.

But for *any* website, sound can be a potent parallel or even alternative information stream, as well as making the site friendlier, more personal, more fun and more accessible (particularly to those who don't read six point Helvetica too well). Think of the engaging way that the Greyworld sonic artists have made the physical world more fun and more interactive by using sound, and then just think how much more can be done in the virtual world, where physics and cost are almost completely removed as barriers to creative imagination.

Before we get carried away, let's remember that the Golden Rules apply here more than anywhere, because the web brings a degree of user freedom that allows for no second chances. When discussing the importance of first impressions and corporate receptions, I mentioned the finding that people assess whether they like the look of a website or not in just 50 milliseconds. While it obviously takes longer to assess a sound, the message is clear. You need to make a good impression and you need to make it quickly. So...

1 **Make the sound optional** – have obvious sound controls; start with sound off; above all, never impose audio that carries on streaming after your website is closed.

2 **Make it appropriate** – consider the audience demographic and the material on the page when you select a sound. Conform to BrandSound™ guidelines if they exist. Also, differentiate between background sound (which simply creates an appropriate ambience) and foreground sound (which relates to a specific item, for example a short quote when you show someone's picture, or the Ferrari engines mentioned above).

3 **Add value** – don't use sound for the sake of it; remember this is a parallel stream. People can look and listen, so don't just read the words on the page – amplify them, give examples, make them come alive, and use people's voices more than anything.

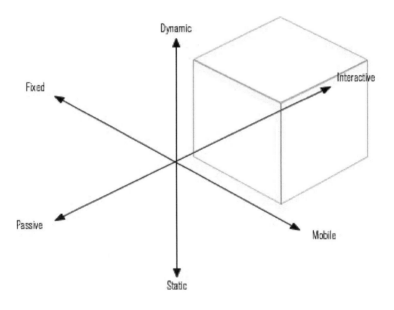

If improved user experiences and increased sales aren't reason enough to add effective sound to your site, how about survival? Consider that there are three dimensions of content, as in the graphic below: first, from passive (the content is fixed or linear, like a novel or film or TV programme) to interactive (the user can change the order, weight, presentation and many other aspects of the content); second, from static (unmoving words and pictures) to dynamic (audio, video and animation); and third, from fixed-access (you have to be still in a certain place to access the content, like TV or cinema) to mobile (you can access the content anywhere, like books, magazines or mobile devices).

I suggest that most of the interesting action in the next decade is going to be taking place in the space marked with the cube: dynamic, interactive, mobile content. For example, suppose I want to know who won the FA Cup in 1958. Which is easier? Stopping what I'm doing, and using my eyes and my hands to operate a screen and keyboard (either fixed or mobile) to make my query and retrieve the answer, or using my voice to query an intelligent agent though my headset and then receiving the answer in audio through the same device?

By using mobile, dynamic, interactive content I am able to keep on doing what I was doing before – walking, driving, making a meal – and achieve the same result. Conversation is our most natural form of communication, and it is inevitable that the web will offer this way of working.

Huge sums of money are being spent on voice input and output, and on the kind of artificial intelligence that will make intelligent agents into virtual personal concierges for us. I believe that within a decade sound will replace keyboards and screens as the primary input/output medium for all our devices. From our work system to our home information/ entertainment system to our personal terminal, we will store, manage and access digital information by talking and listening. Sites like AudioBoo and SoundCloud are leading the way, and doubtless many will follow and create their own rich takes on web sound.

In this new world of the aural Internet, a website that's mute will become invisible too. This means that exploring how sound can enhance your personal, company or brand website is sound business in every sense – but please do use the tools we've explored to make sure that the sound you create is appropriate and effective.

3.10 Telephone sound

According to a survey carried out in the UK by telecoms company Toucan, women spend an average of five years on the phone during their working lives, taking part in around 288,000 calls during that time. Men participate in slightly less calls – around 277,000 – and make shorter calls, with the result that they spend only three years of their lives on the phone. The average business call lasts about five minutes, which is half as long as the average personal call.

Three years or five, that's a huge amount of time spent practising this form of communication. We have become used to doing business on the phone, whether it's checking prices and availability or ordering; arranging meetings or requesting information; cold calling prospects or servicing key customers. Where only a letter would have been appropriate just one generation ago, today a phone call (with possible email confirmation) is standard practice.

The massive increase in telebusiness has some major implications for organisations. First, time has been compressed. An order processing system could have had a turnaround of a week in the 1960s and been considered perfectly satisfactory by customers. Today the same system has to perform in real time* or it will be a fatal wound for the business, haemorrhaging customers with every frustrated call.

Second, the whole process has become more volatile. A 'Dear Sir or Madam' letter was an impersonal communication to a faceless organisation, and the sender would anticipate a similarly emotionally neutral response. A phone call is much more personal, partly because we associate the medium with informality and with personal communication, and partly because conversation transmits emotion so much more powerfully and immediately than written communication. Writing is very susceptible to (mis)interpretation unless used by an expert – hence the rise of crude devices like emoticons in an attempt to limit the huge potential for emails

---

*   This is an interesting euphemism when you think about it: it is taken to mean without any delay at all, but who has decided that this is real time?

and text messages to be emotionally misconstrued.

Even for world-class writers, sound adds power and presence. Think of the difference between reading Shakespeare and hearing it. For the rest of us engaged in commercial communication, the options are web-forms, emails or voice communication. There are probably books to be written about the first two, but since neither of them usually involve sound we will ignore them and focus on voice communication by telephone, including, of course, 'voice over Internet protocol' (VoIP).

When we speak to a 'customer service representative', the voice on the other end of the phone *is* the organisation to us. The accent, timbre, pace, vocabulary, clarity, listening skill, knowledge, attention to detail and most of all the care demonstrated by that person all combine to create our first, and all too often our last, impression of the organisation we are calling. The elements of this call *are* our experience of the organisation and its brand: how fast the call is picked up; how appropriate and effective is the handling of our requirements; the flexibility or lack of it; the style, including music on hold; the respect shown for our needs versus those of the organisation. Every aspect weighs in and shapes our brand experience. Every phone call is an ambassador for the organisation.

Unfortunately the intimacy, immediacy and intensity of the customer experience are too rarely the primary concerns of business, which often sees only one thing in the rise and rise of telephone contact: increased cost. The mathematics are easy to understand. For a software company with a complex product retailing at £50 and a million customers, the words 'customer support' are about as attractive as 'product recall'. The prospect of tens of thousands of lengthy calls a year, each requiring skilled technical staff to answer it, is understandably daunting. With the remorseless march towards 24/7 living, this problem becomes even worse as many organisations face the issue of scalability: how do you meet peak demand without paying for massive redundant resources at off-peak times? And in a global village, how do you cater for the needs of customers in different time zones, all of whom expect instant service?

This economic problem has given rise to an attitude towards customer conversations that fighter pilots would call 'flying inverted'. Any organisation should relish the chance to impress a customer and give superb service. An incoming call requesting product or service information is the best possible opportunity for a sale. Inbound customer calls should be seen as an asset: the more you have, the more your company is worth. But the scalability problem has created the phrase 'call handling', which makes a customer call into a problem to be solved, not an opportunity to be seized – a liability, not an asset. If any niggling doubts remain, the number crunchers can simply aggregate the calls to transform the personal into the general and then the management task becomes one of processing call volumes as efficiently as possible, which means aiming to minimise call lengths and avoid tying up operators with 'trivial' enquiries.

Many organisations have been flying inverted for so long they now think they are the right way up! They don't see the irony in the recorded message that "your call is very important to us" – to which many customers respond with the thought: "Right, so important you can't be bothered to answer it personally."

The typical method for reducing the cost of call handling is two-pronged: first, screen the calls with automated call handling systems so that standard queries get pre-recorded standard answers and only tricky requests get through to the operators; second, outsource the answering of calls to a specialist call centre which can flex personnel to meet varying inbound call volumes.

These methods both process calls at a low primary cost, but what of the secondary cost? In 2005 the on-line insurance firm swiftcover.com carried out a survey of 1,000 people and found that four out of five were stressed and frustrated at trying to contact a call centre. Almost two out of three said they could not understand what the operator was saying – and more than half said they were confused by endless automated options. In the UK every household already spends the equivalent of a whole day on the phone to call centres each year.

The *Hanging on the Telephone* report by the Citizens' Advice Bureau published in 2004 confirmed what we all know anecdotally: people find call centres irritating. The report found that 97 per cent of people surveyed said they found at least one aspect of using a call centre annoying. The most common complaint was being kept on hold for long periods and there was widespread annoyance at being given a multitude of options, and then receiving unsatisfactory service. In this report, perceived standards of service in call centres are shown to be polarised with a significant minority of users dissatisfied with the service they receive.

Satisfaction tends to be highest with call centres operated by industries traditionally committed to delivering good customer service: retailers' call centres generate highest satisfaction, while call centres operated by utilities companies are seen to be least customer-friendly (67 per cent and 49 per cent of customers satisfied respectively). But even at 67 per cent satisfied (probably a result considered good by the organisations concerned) that leaves one third of customers dissatisfied, which should be seen as a disaster! Remember, the received wisdom is that it costs five times more to find a new customer than it costs to retain an existing one.

If there really is no alternative to automation and outsourcing, let's consider some basic principles to get those satisfaction ratings much closer to 100 per cent. First, let's look at automation.

Case study: City of Amsterdam

Tools: audit, sonic logo, brand music, telephone sound,
soundscape, advertising sound

The city of Amsterdam wanted a sound brand for its call centre and digital counter that would be unique and yet immediately recognisable when citizens had contact with these key touch points. The sound had to express the city's core values (active, open, honest).

The city engaged local sound branding agency MassiveMusic, who adopted the keywords 'welcome' and 'trusted' to guide the creative treatment, and embarked on an audit on their bicycles, listening for and recording all the characteristic sounds of the city. Instead of creating a flat collage of sound effects, the agency then embedded the city sounds inside a brand music treatment with a strong modern theme. Massive called this the 'brand score', and also wove into it a sonic logo for the city.

The final treatment was chosen through a classic creative process of mood boards and filtering, and it was then adapted for various uses, including a trance version for the 2010 Gay Pride Canal Parade, ringtones for city employees and soundtracks for advertising.

Customer-friendly automation

1 Always offer a human alternative

These days many people are accustomed to automation and accept it as long as it's effective. But there will always be some customers – for example the elderly, an ever-more important group given the ageing populations of most Western economies – who just want to talk to a person from the outset. Give your customers the choice and serve them the way they want to be served: offer a menu option to go into a queue for a real person in *every* menu you create. If the queue's too long for them, they can call again or return to automation.

2 Always offer an 'other' option

One of the most commonly cited frustrations with menu systems is that the options offered do not cover the customer's need. Many of these systems are poorly designed, and it's vital to make yours as clear and comprehensive as possible, but no matter how good it is, there will always be someone who has a square enquiry where you have offered only round options. Always offer a catch-all 'other' option to give these people somewhere to go (usually straight to a real person); also, log the types of calls that come through this route so that you monitor whether you have missed an important category in your menu.

3 Always offer reverse

It is simply maddening to find yourself in a menu dead end with no way of getting out. Imagine how frustrating the web would be with no back button! It's no different for automated menus, so always offer a key to reverse, and ideally another one to go back to the entry menu. Keep these commands consistent throughout the call.

4 State estimated queuing time

If people opt to go into a queue for a real person, use the technology that's available today to estimate their waiting time and then tell them so that they can decide whether to continue. This will be seen as a fair contract, leaving control of the call firmly where it should be, which is with the customer. Blind queuing is guaranteed to create stress and frustration, so that one group of customers will just leave in a funk (and probably never come back) and the rest will become angry and resentful, with the consequence that operators will have to defuse callers' upsets before they can start to talk business. That wastes time and stresses operators, making them less effective and increasing the already high burnout rates in this job.

5 Offer call back

When offering the option to queue and estimating the queue length, give people the option to be called back. This is a quantum improvement in customer service but it's rarely deployed as yet. Organisations offering this option identify themselves as really concerned with customer service, and will gain by experiencing much lower attrition rates and generating more business in shorter, less stressful calls.

6 Never charge for waiting

Some automated systems add insult to injury by forcing people into lengthy, blind waiting – and charging them for the privilege. It would be insulting to charge people to stand in a physical queue, and it should be equally inconceivable to treat callers this way. It may be acceptable to offer the option of paying a small charge for faster response – again, that would be putting the choice with the customer and then it would simply be a question of getting the price point right so that a manageable number choose this option. As always, the rule is to give the customers choices, rather than to herd them around like sheep.

7 Consider style as well as content

The voice or voices used, the way they deliver their lines, the on-hold sound – all these are expressions of your organisation. So it's curious that many automated systems use obviously untrained voices, stiffly and uncomfortably reading from a poorly-written, unnatural script. At the other extreme, many other systems use actors who seem to have been briefed to sound as smug, overpowering and insincere as possible. Getting the right balance takes care but this is not a new art – ask any radio station or TV voice-over production facility. It's time we moved the automated call system up to match the standards of auditory presentation customers are used to in these areas. Engage professionals to design and deliver the audio so that it lives and breathes your brand. Offer people a choice of soundscapes while they hold – various styles of music and some natural sound alternatives such as birdsong or surf would at least emphasise that you know that not everyone likes Candy Dulfer. There are some very good and dedicated companies now focusing on creating custom on-hold sound. I have worked with Manchester-based PH Media, just one such company, and have been impressed by their professionalism and imagination. Their research shows that (a) most companies have poor on-hold sound and are damaging themselves as a result (unknown to the top management, who rarely if ever call themselves to have the experience their customers do); and (b) significant sales uplifts can result from well-designed on-hold sound. This is an investment with a high return.

8 Continually optimise

This is the front line of customer service for any organisation, and yet many create their automated systems and then just leave them to run unchecked for months or even years. It is vital to plan and carry out continual optimization and maintenance. In the first place, you need to know the real cost of your system (including lost customers and damaged reputation) in order to make correct decisions about future investment in it. In the second, the world changes rapidly and no system is ever optimal so you must continually research, secret shop, test and test again. Every aspect needs to be challenged regularly, tested with real customers and optimised for effectiveness, consistency, speed, common sense, flexibility and style.

9 Be straight

People spot insincerity fast, and they don't like it. Please don't tell your customers how important their call is if you're actually doing your best not to talk to them – the result will be a reputation holed below the waterline. If you get a lot of calls and you can't physically employ enough operators to deal with them, why not tell people the truth? They will respect you for it and at least they will know where they stand. Let's swap the fantasy world of Customer Service Representatives for a real one of automation for mutual benefit and tiers of service where customers can choose the combination of cost, speed and quality that is ideal for them.

Hopefully in time we will arrive at a consistent user interface that works the same way in every call – much like the way we all know how to deal with a website. If more organisations follow these guidelines, there is no reason why using an automated system should not be a positive, even enjoyable experience.

The call centre

With customer-facing automation in place we now have less angry and upset people, less terminated calls, less lost business. But many of the callers still want to talk to a real person. And so we move on to the call centre.

For any organisation experiencing high inbound call volumes, the logic of centralising resources to focus training, to maintain consistency and to achieve economies of scale with hardware and software is crystal clear. As before, the problems start to occur when management shifts from seeing calls as opportunities to win lasting customers, towards seeing them as burdens to be dealt with as cheaply and quickly as possible. The end result is truly flying inverted: nobody in the organisation has any contact with customers at all. This is the epitome of the customer-as-nuisance view: if only those pesky customers would stop bothering us we could get on with doing our jobs!

Whatever the philosophy, the economics are impossible to dispute, so centralised call handling is a huge and fast-growing business. In the USA, over three million people work as customer service representatives, while in the UK the total is over a million, and the industry continues to grow.

The most rapid growth, however, is in overseas call centres, the management-speak for which is 'Offshore BPO'. At first dominated by English-speaking economies such as Canada, Australia, New Zealand and South Africa, we are now seeing a boom in the Asia-Pacific region, where countries like India, China, the Philippines and Malaysia boast industries that have grown from nothing ten years ago to compete with any in the world. Growth rates are typically over 40 per cent a year, sustained over long periods. Labour costs are as little as one tenth of those in the older economies, so total call handling costs are typically one fifth.

This is a young industry and there are signs of structural weakness: the staff attrition rates are very high – from 25 per cent to over 100 per cent per annum – and there have been instances of fraud, as well as accusations of exploitation of 'cyber coolies'. Certainly recruitment, training, quality control and motivation are major challenges with the industry's insatiable demand for more and more people.

However this industry deals with its challenges, it is clear that call centres, whether in-house, locally outsourced or overseas, have a major role to play in solving every organisation's problem: how to satisfy mass, often global, customer demand for instant telephone service, 24/7.

What we need to be certain of is that customer calls are never 'out of sight, out of mind'. The following guidelines for operating customer-friendly call centres should make sure that never happens.

Customer-friendly call centres

1 Never compromise quality

Let call quality be your one fixed criterion. Benchmark it quantitatively (number of rings to answer, percentage of queries completely satisfied, speed of response, minimal in-call holding and so on). Never compromise on it. If you can buy it more cheaply outside or overseas, fantastic. But don't accept low quality to get low prices because your business will suffer in the long term.

2 Choose a reputable operator

Look for membership of reputable trade associations with standards and dispute resolution procedures. Also consider length of trading and client lists. Most importantly, ask for and take up references.

3 Install local management

Whatever promises are made, call centre staff do not work for you. Even if they care as much as you do about your reputation (unlikely) they can't know everything your own staff do – the history, background, personalities, myths, stories and culture are missing for them. The best way of minimizing the effects of this gap is to install in the call centre at least one person who *does* have all this, to work with a consistent team, monitoring calls, training continuously and motivating them to care.

4 Define the style as well as the content

Often what is said is not as important as how it's said. There are several factors to consider here, all of them made so much easier to control if you have BrandSound™ guidelines that cover brand voice on the telephone.

Is there a gender or age profile that's appropriate for your brand? For example, Tampax should probably have all women in their call centre; Nickelodeon should have young voices.

Is there a brand voice style? For example, a book retailer like Waterstones might sound calm and intellectual; Pepsi Max might be high-energy and vibrant. Choose people who fit the profile and train the team to deliver it consistently.

Is there a technical qualification? Nestlé in Paris had a pet food call centre where all the staff were qualified vets. For any technical support call centre, it's vital that the staff are trained and able to go off-piste and think on their feet, treating each customer/problem as an individual and carrying out synthetic diagnosis – not just reading scripts from a database.

If you're outsourcing abroad, know what's acceptable in terms of strength of local accent, be it Irish, Scottish or Indian, and interview team members to ensure you maintain that standard.

Finally, consider the on-hold sound and choose something that is appropriate and that adds value, either by being pleasant or by delivering useful information. BSkyB's tech support on-hold sound is a recorded announcement giving simple answers for their most commonly requested user problems, which means that many callers don't have to wait to speak to a real person.

5 Use scripts as a fall back, not a first resort

After years of the conventional wisdom that a rigid script is essential for quality control, the tide has at last started to change. UK bank Lloyds TSB carried out research and found that nine out of ten people were annoyed by obvious script use, that 60 per cent feel their questions are not answered when scripts are used, and that 55 per cent feel that scripts prevent agents from really listening to them. To cap this, 86 per cent of the agents were also in favour of getting rid of scripts. As a result, the bank's UK-based call centres have dropped scripts, joining their overseas call centres, which had never used them. Though there are times when scripts are useful (for example for arcane, rarely-asked questions or where precise wording is critical) they can never be a substitute for an authentic dialogue with a well-trained, well-informed person. Please retreat to them only if necessary, rather than deploying them by rote.

6 Do not emphasise speed to clear

This is probably the statistic that most reflects the flying inverted attitude to customer calls. Obviously there will be some calls that outstay their welcome, but how many anecdotes from management gurus can we all remember about patience and unswerving dedication to service on the telephone yielding amazing results? Or about brusqueness and prioritising the organisation's own agenda over those of the customers leading to losses, not to increased efficiency and profit?

Why would we want to talk to our customers for the shortest possible time? One-to-one time with a customer is a precious asset and a huge opportunity. Please don't throw this baby out with the bathwater – it is the future of your business, because if nourished and cherished it grows into massive customer loyalty, which other departments are probably already spending millions trying to generate.

7 Check, check and check again

Employ an independent organisation to check that the defined standards are being maintained at all times by secret shopping against a well-defined checklist and plan of call types. Double check these metrics against the classic ones your call centre will be giving you.

Case study: UK broadband supplier

This is a perfect example of how to upset a customer with the upside-down attitude that calls are a nuisance dictating the way the whole system is set up. The objective of my call was to check whether a requested upgrade to a broadband service had been carried out as promised, which I suspected had not been done. It turned into a modern-day version of the classic Flanders and Swann song The Gasman Cometh.

1 Billings.

Three-minute hold. Pleasant man confirms that the account has not been upgraded, gives me an 0845 number to call.

2 0845 number.

Just two recorded announcements, both of which are dead ends with no way of backtracking; one of them is cancel your account! There is no option to speak to a human. Dig out the customer care number.

3 0870 customer care number.

Automated menu system has six options, none of them appropriate, none of them to speak to a human. In an effort to get through to a real person I choose 'change address'. This takes me to a long recorded message telling me how to update my address on the website, with no further options and no way of backtracking. After multiple calls I finally find a way to opt to speak to a human, who will be with me 'shortly'.

10-minute hold.

Finally picked up by Sheba in the customer care call centre in Bangladesh. Strong regional accent. Calls me Mr Taylor throughout, enters the wrong phone number to check, cannot understand my questions, makes incorrect assumptions about the upgrade request...

I ask for a supervisor or for the phone number of the company in the UK. They have no number for the company in the UK so she will put me through to a supervisor.

<u>*Four-minute hold with no sound at all.*</u>

Have I been cut off?

Sheba comes back to say the supervisor is here now.

<u>*Silence for two more minutes.*</u>

The supervisor finally picks up; says I have been calling the wrong number and gives me a different 0870 number to call. I have been on this call for 23 minutes.

4 New 0870 number (tech support I think).

Automated menu system offers several options, none of which apply, with no option to speak to a real person. Finally fight through the system to a representative in a New Delhi-based call centre. I give him the details all over again. He is very polite and he goes to check.

Four-minute hold with cool, smooth music. This is not how I feel.

He comes back and confirms the upgrade was not done, and that I should get a credit. He will action the upgrade but he can't do the credit.

The department I need to call is Billings…

Part 3 References

79 Martin Lindstrom (2005) BRANDsense, Kogan Page, 2005

80 Michael Bull and Les Black, The Auditory Culture Reader, Berg, 2004, p2

81 Corporate Sound als Instrument der Markenführung, January 2006, available from www. metadesign.de

82 Bertil Hultén, Niklas Broweus, Marcus van Dijk (2009) Sensory Marketing Palgrave Macmillan

83 Passikoff and Shea (2008), The Certainty Principle, Authorhouse, p 11

84 for example Bronner, Hirt, Ringe (2010) Audio Branding Academy Yearbook 2009/2010, Intl Specialized Book Service Inc

85 Michael Harris Cohen, Michael H. Cohen, James P. Giangola, Jennifer Balogh (2004) Voice User Interface Design, Addison-Wesley, p 76

86 Quoted by Ruth Simmons in What Every CMO Should Know About Sound and Music, published in Audio Branding Academy Yearbook 2009/2010, Nomos, p92

87 Dan Jackson (2003) Sonic Branding: An Essential Guide to the Art and Science of Sonic Branding, Palgrave Macmillan

88 Zentner, Marcel, Didier Grandjean, and Klaus R Scherer. Emotions Evoked by the Sound of Music: Characterization, Classification, and Measurement. Emotion 8, no. 4 (2008)

89 www.radioadlab.org

90 J. Alpert and M. Alpert (1989) Background music as an influence in consumer mood and advertising responses, published in Advances in Consumer Research Vol 16, pp 485-491.

91 Park, C. W. & Young, S.M. (1986). Consumer Response to Television Commercials: The Impact of Involvement and Background Music on Brand Attitude Formation published in Journal of Marketing Research, Vol. XXIII (February), 11-24.

92 J. D. Morris & M. A. Boone, (1998) The Effects of Music on Emotional Response, Brand Attitude, and Purchase Intent in an Emotional Advertising Condition, published in Advances in Consumer Research Volume 25, pp 518-526.

93 Please follow the link on the book's website to see the scans of this article on the Modern Mechanix blog.

94 Chion, M. (1994) Audio-Vision, Columbia University Press p 34.

95 Lindstrom, ibid, pp 12 and 50

96 Fidell S (1978) Effectiveness of audible warning signals for emergency vehicles. Human Factors, 20:19-26.

97 www.delta.dk

98 Lindgaard G., Fernandes G. J., Dudek C. & Brown, J. (2006) Attention web designers: You have 50 milliseconds to make a good first impression! published in Behaviour and Information Technology, vol 25 pp 115-126. The authors tested subjects by displaying web pages (saved to disk) for 500ms or 50ms for "visual appeal." The report concludes that "...visual appeal can be assessed within 50 msec suggesting that web designers have about 50msec to make a good first impression."

99 http://news.com.au 5 Oct 2004

100 See http://nomuzak.co.uk or www.pipedown.info as examples – there are others.

101 Gavin, Helen (2006) Intrusive music: the perception of everyday music explored by diaries in The Qualitative Report September 1, 2006

102 Adrian North quoted in The Independent article Suicide link to D-I-V-O-R-C-E by Charles Arthur, London, September 12, 1996

103 In-Store Magazine, February 2006, p27

104 Ananova 3 Dec 2003

105 North, A. C. and Hargreaves, D. J. (1998) The effect of music on atmosphere and purchase intentions in a cafeteria. Journal of Applied Social Psychology, 28, 2254- 2273 and also Wilson, S. (2003) The effect of music on perceived atmosphere and purchase intentions in a restaurant. Psychology of Music, 31, 93-109.

106 Kjellberg, A., Landstrom, U., Tesarz, M., and Soderberg, L. et al. (1996) The effects of nonphysical noise characteristics, ongoing task and noise sensitivity on annoyanceand distraction due to noise at work. Journal of Environmental Psychology, 16, 123- 136.

107 Klitzman, S. and Stellman, J. M. (1989) The impact of the physical environment on the psychological well-being of office workers, published in Social Science & Medicine, 29, 733-742.

108 Knez, I. and Hygge, S. (2002) Irrelevant speech and indoor lighting: Effects of cognitive performance and self-reported affect, published in Applied Cognitive Psychology, 16, 709-718.

109 Evans, G. W. and Johnson, D. (2000) Stress and open-office noise. Journal of Applied Psychology, 85, 779-783.

110 Hedge, A. (1982) The open-plan office: A systematic investigation of employee reactions to their work environment. Environment and Behavior, 14, 519-542.

111 Jackson, T. S. (1999) Irrelevant speech, verbal task performance, and focused attention: A laboratory examination of the performance dynamics of open-plan offices (mental workload, work performance, computer use, noise). Dissertation Abstracts International: Section B: The Sciences & Engineering, 60(6-B), 2997.

112 Loewen, L. J. and Suedfeld, P. (1992) Cognitive and arousal effects of masking office noise. Environment & Behavior, 24, 381-395.

113 Willner, P. and Neiva, J. (1986) Brief exposure to uncontrollable but not to controllable noise biases the retrieval of information from memory. British Journal of Clinical Psychology, 25, 93-100.

114 Young, H. H. and Berry, G. L. (1979) The impact of environment on the productivity attitudes of intellectually challenged office workers. Human Factors, 21, 399-407.

115 Banbury and Berry, ibid

116 Karageorghis, C.I. & Terry, P.C. (1997) The Psychophysical Effects of Music in Sport and Exercise: A Review, Journal of Sport Behaviour, 20(1), 54-68

117 For example see Ros Bandt (2005) Designing Sound in Public Space in Australia: a comparative study based on the Australian Sound Design Project's online gallery and database Organised Sound, 10: 129-140 Cambridge University Press

118 Bronzaft, A.I. and McCarthy, D.P. (1975) The effect of elevated train noise on reading ability, published in Environmental Behavior, 7, 517-528. 15; also Evans, G.W., and Maxwell, L (1997), Chronic noise exposure and reading deficits: The mediating effects of language acquisition, published in Environment and Behavior 29(5), 638-656. 13.

119 E Durham, (1991) Relaxation Therapy Works, RN Aug 91, pp 40-42

120 Hospital Noise — Levels and Potential Health Hazards, Stephen A. Falk, M.D., and Nancy F. Woods, R.N., M.N., in N Engl J Med 1973; 289:774-781 October 11, 1973

121 Quick, Brian. A Longitudinal Study Examining The Priming Effects of Music on Driving Anger, State Anger, and Negative-Valence Thoughts Paper presented at the annual meeting of the International Communication Association, Marriott Hotel, San Diego, CA, May 27, 2003

122 Household Penetration Rates for Technology Across the Digital Divide, Tomi Ahonen blog Jan 20, 2011 at http://communities-dominate.blogs.com

123 The Independent, May 29, 2006

124 A.S. Niskar, S.M. Kiesak, A. Holmes, E. Esteban, C. Rubin and D.C. Brody (1998) Prevalence of Hearing Loss Among Children 6 to 19 Years of Age published in Journal of the American Medical Association, vol. 278, pp 1071-1075.

Postscript: the future

I sincerely hope that this book will help to open a door that's been closed for more than two hundred years, allowing greater access to a world of careful listening and well-designed, conscious, intentional, applied sound. I have seen the level of interest in sound grow greatly over the last few years. People find it fascinating, and with good reason because it affects them so much. I hope that *Sound Business* will contribute to the growing debate, and will act as a guide, a reference work and a practical handbook for the people who are responsible for most of the world's sound: the managers of private and public sector organisations.

We have created a website to accompany this book at www. soundbusiness.biz. As well as links to download all the sound samples flagged in the book (these are all hosted on SoundCloud), it has links to my blog, AudioBoo and Twitter, to a Squidoo lens on sound and also to my TED talk on the four effects of sound, as well as to The Sound Agency's YouTube channel and SlideShare page. You can also sign up for a regular e-newsletter, which is called *Sounds Interesting*. I invite you to visit the site and subscribe to the newsletter to keep abreast of developments in the world of sound for business.

I warmly welcome comments, suggestions, experiences and other communication about sound in business. You can contact me via the book's website or at julian.treasure@thesoundagency.com. I am also on LinkedIn, Ecademy and Xing if you want to connect through those networks.

I look forward to hearing from you.

The sound files

Visit www.soundbusiness.biz and you will find links to the sounds listed below, and more posted since this book's publication.

Topic	Track	Name	Notes
Vibration	1	CMB	Cosmic microwave background: the sound of the Big Bang, transposed to audible frequencies (as are the other cosmic sounds*)
	2	Black hole	The sound of X-rays escaping in jets from Black Hole GRS 1915+105
	3	Pulsar 1	PSR B0329+54
	4	Pulsar 2	Vega PSR B0833-45 – not a bad drummer!
Sound	5	Sax, drum, trumpet violin	To accompany waveform graphics
	6	Sine waves	All the audible A notes from 27.5 Hz to 14.08 kHz
Acoustics	7	Supermarket café	A quiet cuppa? I think not.
	8	Our office	Before and after we treated the walls
	9	Retail café	Before and after we treated the ceiling
WWB	10	Wind	In leaves and grass
	11	Water	Rain, surf, stream, lake
	12	Birds	In various locations, then a slowed down section

---

* These amazing sounds are all taken from the websitewebsite www.spacesounds.com with thanks to its creator Scott McNulty. If you like this kind of thing, there is more there, and also at www-pw.physics.uiowa.edu/space-audio/sounds/.

Generative sound	13	BAA1	Both birds and musical elements are generative
	14	Chemistry	A short snapshot – the real thing is live if you call !
	15	Shetland	A collage of the three scenes we created, live at the museum every day
Noise	16	White, pink, grey	
Beat frequencies	17	Beat frequencies	On headphones you can hear the beat frequency as the two tones are panned together into the centre
	18	Cerego natural soundscape with beats	
Sonic logos	19	Intel, Hemglass, Apple, MGM	
Product sound	20	Bacon	A sizzling opportunity
	21	Chiller cab mix	
	22	Coffee machine	
	23	Diesels	
Soundscapes to avoid	24	Shops	A mix of several high street shops
	25	Supermarkets	A mix of several supermarkets
	26	Service trolley	Over 80 dB at one metre as it passed
	27	Hotel	Hey Jude, don't make it *this* bad!
	28	Office	Typical offices – as usual, listening without visuals emphasises the noise
	29	Transport	A mix of typical transport soundscapes. Prize to anyone who can understand the PA announcement

	30	BOX	A collage of sounds we created for BOX
Soundscapes to enjoy	31	Water/harp	Loopable sound we created for an analgesic, who didn't use it – so you can. Set 'repeat track' on your player and enjoy this one and the remaining three tracks for long periods.
	32	BAA day	12 minutes long; perfect for working
	33	BAA night	12 minutes long; ideal for resting
	34	Lake birds	We use this in our office all day

Further explorations

Below are a few of the books I have enjoyed or found particularly useful. Many (but not all) of them appear in the references section. For further resources I recommend the CAIRSS website, a searchable archive of thousands of relevant papers and journals, to be found at http://imr. utsa.edu/CAIRSS.html, and for connections with people interested in sound branding try the Audio Branding Academy website at http:// audio-branding-academy.org/, within which you can also find the International Community for Audio Branding.

Joachim-Ernst Berendt *Nada Brahma: The World is Sound* (Destiny Books, 1991)

Barry Blesser *Spaces Speak, Are You Listening? Experiencing Aural Architecture* (MIT Press, 2009)

Bronner, K., Hirt, R. and Ringe, C. (2010): *Audio Branding Yearbook 2009/2010*. Nomos Edition Fischer, Baden-Baden.

Bronner, K.; Hirt, R. (2009): *Audio Branding. Brands, Sound and Communication*. Nomos Edition Fischer, Baden-Baden.

Donald Hodges (ed) *Handbook of Music Psychology* (IMR Press, 1996)

Hazrat Inayat Khan *The Mysticism of Sound and Music* (Shambhala, 1996)

Bart Kosko *Noise* (Viking, 2006)

Joshua Leeds *The Power of Sound* (Healing Arts Press, 2001)

Martin Lindstrom *BRANDsense* (Kogan Page, 2005)

David Rothenberg and Martha Ulvaeus (eds) *The Book of Music and Nature* (Wesleyan University Press, 2001)

R Murray Schafer *The Soundscape: Our Sonic Environment and the Tuning of the World* (Destiny Books, 1994)

Steven Strogatz *Sync: The Emerging Science of Spontaneous Order* (Theia Books, 2003)

David Toop *Haunted Weather: Music, Silence and Memory* (Serpent's Tail, 2004)

Barry Truax *Acoustic Communication* (Ablex Publishing, 2001)

To help understand and practice conscious listening and conscious sound creation I recommend two wonderful, award-winning DVDs.

Evelyn Glennie *Touch the Sound* (Docurama, 2006, directed by Thomas Riedelsheimer)
Stomp *Out Loud* (VCI, 1999)

Index